WATER IN THE MIDDLE EAST
COOPERATION AND TECHNOLOGICAL SOLUTIONS IN THE JORDAN VALLEY

Studies in Peace Politics in the Middle East

Volumes 1, 2, and 4 are published in association with the
University of Oklahoma Press

Contents

Foreword

HRH PRINCE EL HASSAN BIN TALAL
President of the Club of Rome
Moderator of the World Conference on Religion and Peace
Chairman of the Arab Thought Forum

Like the banks of the Nile, the Euphrates, and the Tigris, the Jordan River Valley bore witness to the advent of human civilization as early peoples exchanged pastoral nomadism for settled life in agricultural communities. Recent archeological evidence indicates that the waters of the Jordan River basin have been used for irrigated agriculture for millennia – as far back as the Stone Age.

While the Nile has played a sustained role in the development of human civilization, owing to the annual flooding of its banks, the Jordan River and Tigris–Euphrates basins have been more dependent upon human efforts, particularly the irrigation systems constructed and cared for by early states. Fluctuations in the power and progress of regimes were paralleled by the improvement or deterioration of the irrigation infrastructure, which affected food production, for better or for ill, and the rise and fall of states.

Throughout most of recorded history and until 1920, the Jordan River basin was under the jurisdiction of only a single sovereign power at any given time. Today, the basin straddles four states – Lebanon, Syria, Jordan, and Israel – to which a fifth may soon be added – Palestine. It thus became imperative, during the twentieth century, to envision and establish some unified plan for the basin's development so that projects implemented in downstream riparian countries would not be negated by others in upstream countries. Given the region's political environment and the fact that the relevant Arab states were in a technical state of war with Israel, the United States agreed to use its good offices to attempt to devise a unified plan for the development of the Jordan basin. An American presidential envoy began a series of consultations with each of the riparian countries, traveling between them and Egypt in what would later be known as shuttle diplomacy, and finally secured the consent of all the parties at the technical level. In October 1955, the Unified or Johnston Plan for the development of the Jordan basin was formalized; in January of the following year, it was distributed to the parties. More time was needed, however, before agreement was possible at the political level and ratification could take place.

So long as all of the signatory states adhered to its provisions, the plan sustained riparian and other development in the Jordan Valley. In 1959, Jordan drew no more than its rightful share of the waters of the Yarmouk River when initiating its East Ghors Canal Project; and Israel did the same when constructing the Tiberias–Beit Shean pipeline to irrigate areas in the western Jordan valley. As Jordan expanded its Jordan Valley Project, Israel also increased its use of the basin's waters, pumping more of them into Israel's National Water Carrier. States upstream, like Syria and Lebanon, have not always abided strictly by the agreement. Syria exceeded its water-sharing quotas as stipulated in the Unified Plan and, between 1967 and 1987, systematically extracted more water from the Yarmouk than it was entitled to; this had a negative impact upon the implementation of Amman's plans for the eastern Jordan Valley. The most recent disagreement erupted in 2002, when Israel contested a Lebanese project to draw water from the Wazzani springs (feeding the Hasbani tributary) for domestic consumption in some adjacent villages. The United States was asked to intervene to calm the situation and Lebanon proceeded with the project a short time later.

Water-sharing among states highlights the difference between times of war and times of peace and nowhere has this contrast been more evident than in the Jordan Valley. The 1967 war and the subsequent fierce confrontation between Israel and paramilitary and military formations in the eastern Jordan Valley brought destruction and devastation in their wake. Over 60 percent of houses in the Jordan Valley were destroyed. Virtually all of its population fled to take refuge in the safer eastern plateau. As law and order were re-established, I had the privilege of overseeing the planning and implementation of projects for the rehabilitation and development of the eastern bank of the Jordan Valley, a process that I supervised with great interest for over a decade and a half (1972-1988). With the help of friendly countries, institutions, and organizations from around the world, Jordan was able to make the Jordan Valley what it is today. The backbone of the development effort has been water from the Yarmouk and from wadis on the valley's eastern side.

The contrast between war and peace is indeed staggering. From about 63,000 people in 1972, the valley's population has grown to over 250,000 today – an average growth rate of over 6 percent per annum. Half of these people were attracted to the valley by new job opportunities and improvements in the physical and social infrastructure. The per capita share of national income in the valley skyrocketed from about US $250 per annum to about $1600 in a decade and a half, paralleling the increase in per capita income countrywide. The rural Jordan Valley became the frontline of the kingdom's overall development and a success story that invoked pride in the contributing agencies, the Jordanian government, and, of course, Jordan's people.

As a Jordanian, I take pride in the fact that the Jordan River Valley now encompasses the Jordan Rift, between the Yarmouk and Aqaba, owing to the addition of the territories of the southern Ghors and Wadi Araba. Development has also been impressive in these added regions, especially in the southern Ghors, adjacent to the Dead Sea.

Yet, today, despite all that has been done, the need for the coordinated development of the Jordan Valley is perhaps greater than before. During peace negotiations between Jordan and Israel, I took my place next to my brother, the late King Hussein; one of my many duties was to supervise the preparation and presentation of a document entitled "Integrated Development of the Jordan Rift Valley." This plan was adopted by the Trilateral Economic Committee (made up of Jordan, Israel, and the United States) and a pre-feasibility study of it has been completed. It envisions transferring sea water from the Gulf of Aqaba to the Dead Sea in order to raise its depleted levels. Part of that same flow would be desalinated, while also generating the modest amount of power needed to pump the desalinated water to consumption centers in Jordan, Israel, and Palestine. The attractiveness of this plan is augmented by the expectation that it will lead to improvements in the valley's infrastructure and environment. Once it is implemented and operational, it will form a venue for sustained cooperation between the three states.

Water, especially the waters of the Jordan basin, can play a significant role in building peace and ensuring that riparian neighbors work together to achieve mutually beneficial goals. Water, so essential to life, is the means to extinguish fires, not to ignite them. I have worked over the years to promote the formation of a new framework for cooperation in the greater Middle East and North Africa (MENA) – a "Community of Water and Energy" similar to the Community of Coal and Steel that was established in Europe in 1951 and served as a precursor to the European Common Market. The ability of Europe to overcome its troubled past and achieve full-fledged union in only half a century is an example to inspire us all. Water and energy have a unique synergy and deserve to be developed and managed in an integrated fashion. Yet, while some of the region's countries are endowed with water resources and others with energy resources, few are endowed with both. Perhaps the countries of the MENA region can embark on steps similar to those taken by Europe to devise an effective system of cooperation. Cooperation among neighbors is crucial to positive outcomes by which all stand to gain.

El Hassan bin Talal

Preface

DAVID L. BOREN
President of the University of Oklahoma

———

Academics and practitioners alike are giving more attention to the relationship between environmental security and global security in international relations and world politics. There is growing conviction among many in academia and the American establishment – notwithstanding the United States' foray into Iraq and the continuing "War on Terrorism" – that as the world's superpower, the United States must give greater priority and resources to become a "war preventer" and a "global stabilizer."

The harsh realities of disorder and violence so preeminent in world politics today are caused by a number of destabilizers. Poverty, famine, disease, a lack of physical human resources, lack of political and social justice, and incapable governance are breeding grounds for a myriad of destabilizers, such as social violence, crime, anarchy, refugees, drug and human trafficking, religious intolerance, insurgency, terrorism, and environmental degradation. These destabilizing conditions are exploited by power seekers motivated by nationalism, religious fundamentalism, ideologies, and power itself. They result in civil wars and inter-state wars and threaten the wellbeing of stable nations.[1]

Thomas Homer-Dixon and Ted Robert Gurr are scholars who in analyzing contemporary conflict have placed emphasis on this kind of cause and effect. Homer-Dixon argues that states' internal instability, which spills over into neighboring states, cannot be dealt with without also dealing with their root environmental causes.[2] Gurr, in *Why Men Rebel*, proposed the concept of "relative deprivation." Working with colleagues in "The State Failure Project," he later argued that relative deprivation may be caused by environmental stress and can lead to internal and external violence and state failure.[3]

This book, *Water in the Middle East: Cooperation and Technological Solutions in the Jordan Valley*, edited by K. David Hambright, F. Jamil Ragep, and Joseph Ginat and published by the University of Oklahoma

Press in its International and Security Affairs Series, makes an important contribution to our understanding of the link between environmental factors and political stability. Because water's attributes make it known as the source of life, the scarcity of water in the Middle East complicates politics and can make finding a solution to the conflict more difficult, especially the Israeli–Palestinian situation. Growing population and economic activity are exacerbating the challenge. In typologies that rank countries on their adequacy of water supplies, Israel and Jordan are placed in categories of "absolute scarcity." The annual per capita water availability in the United States is more than thirty times that of Jordan and more than twenty-five times that of Israel. Per capita water usage of Israelis is five times that of Palestinians.[4]

The issue of allocation and control of water resources can be a source of direct conflict and dispute. On the other hand, due to the shared nature of water, which requires joint management and protection from pollution, there is a strong incentive for cooperation. The availability of water is increasingly considered a humanitarian issue and increasingly described as a "right," so that misallocation of water is regarded as a breach of legal norms. Water, more and more, is a national security issue. Some analysts have argued that while wars of the twentieth century often had causes related to energy resources, those of the twenty-first century may be more likely to originate in contests over water.

In the Israeli–Palestinian conflict, water control and allocation have been and remain disputed matters. Interestingly, most observers have thought that talks on water contribute to progress in peace talks. Whereas cooperation has halted in nearly every area, especially since the start of the *Second Intifada* that began on September 28, 2000, when Ariel Sharon visited the Temple Mount, cooperation on water has continued. The only agreements reached by the government of Israel and the Palestinian Authority during the *Second Intifada* have both involved water. Water has been considered to be so important that it should not be politicized. There has been a view to this time that water is so essential and symbolically important that dialogue on water can contribute to rapport and trust that can enhance the negotiating environment for a final peace arrangement.

The secretariat of the Center for Peace Studies is housed in the University of Oklahoma's International Programs Center. The center is constituted by the University of Haifa (Israel), Bethlehem University (Palestine), the Horizon Institute (Jordan), and the University of Oklahoma (United States). The center contributed to Middle East dialogue on water by holding a conference on the subject at the University of Oklahoma in November 2001. It was attended by outstanding water experts from the region. During this conference the Middle East Water Working Groups (MEWWG) project was conceived, and, with funding from the Bureau of

Educational and Cultural Affairs of the U.S. Department of State and the University of Oklahoma, the project is being implemented. The 2001 conference is described in the *Introduction,* and the functioning of the MEWWG is described in the *Postscript.* This volume is a product of the conference and the project. A total of ten workshops have been held with water expert participants from Jordan, Syria, Turkey, Iraq, Palestine, and Israel. The book is the third volume of the University of Oklahoma Press's International and Security Affairs series. It moves the series into a new area – the examination of resources and ecology as a source of conflict. Publication of this book comes at a critical and dangerous time in the Middle East.

Water talks can contribute to overall peace negotiations, and they can also be a point of dispute. There are some who believe that the water situation for Palestinians must begin to improve notably, especially in terms of equitable treatment with Israelis, or water may become a catalytic element for Palestinian protests and violence. Part of this derives from second track diplomacy fatigue after years of no negotiations.

This book provides provocative, innovative, and practical scientific analyses that should be considered in solving the technical, political, and economic problems related to water division, scarcity, and pollution. The political and economic costs of delaying settlement of water problems through technical solutions and regional cooperation will be vastly greater than the current costs attached to such solutions. This book provides information and possible solutions that should be studied and weighed by all parties involved in Middle East peace negotiations.

Notes

1   Frank McNiel and Max G. Manwaring, "Introduction: Making Sense of Environmental Security" in Max G. Manwaring, editor, *Environmental Security and Global Stability: Problems and Responses* (Lanham, MD: Lexington Books, 2002), p. 5. See also Leslie Gelb, "Quelling the Teacup Wars," *Foreign Affairs* (November–December 1994), p. 5.

2   Thomas Homer-Dixon, "On the Threshold: Environmental Changes as Causes of Acute Conflict," *International Security* (1991), pp. 76–116; and *Environment, Security and Violence* (Princeton, NJ: Princeton University Press, 1999).

3   Ted Robert Gurr, *Why Men Rebel* (Princeton, NJ: Princeton University Press, 1970); Ted Robert Gurr, "On the Political Consequences of Scarcity and Economic Decline," *International Studies Quarterly* (1985), pp. 51–75; and John L. Davier and Ted Robert Gurr, editors, *Preventive Measures: Building Risk Assessment and Crisis Early Warning Systems* (Lanham, MD: Rowman & Littlefield, 1996).

4   World Resources Institute, *World Resources, 1996–97* Databases (Washington, D.C.); and Melissa Fair, "Working Groups to Study Water Resources and Usage in the Middle East, Southern Tier Meeting 3: Israelis,

Jordanians, and Palestinians." A report prepared on the Aqaba, Jordan, June 25–26, 2004 meeting for the University of Oklahoma, International Programs Center, Center for Peace Studies, July 2004, pp. 12–15.

Acknowledgments

The editors are indebted to many individuals and institutions for making this book possible. The Foundation for Environmental Security and Sustainability (FESS) provided funding for the November 2001 conference upon which this book is based. Julie Horn, Kathy Shahan, Gary Miller, Donna Cline, Patsy Broadway, Ragan Traughber, and the late Loretta Selvey of the International Programs Center at the University of Oklahoma gave invaluable support throughout this project. We are also grateful to the staff of the University of Oklahoma Press and Sussex Academic Press, especially John Drayton and Patsy Willcox (UOP) and Anthony Grahame (SAP), for all their efforts. Pamela Genova and Sally Ragep provided much needed editorial help. Ambassadors Edward J. Perkins and Edwin G. Corr imparted the vision and inspiration that made this foray into the torrent of Middle Eastern water issues possible. We are especially beholden to Ambassador Corr, the editor of the series in which this book is a part, who was unceasing in his efforts to have this volume appear. Finally, we must thank the conference participants and contributors themselves, who have inspired us with their determination to transcend disciplinary, national, ethnic, and religious boundaries in the quest for meaningful dialogue.

CHAPTER 1

Introduction:
Water – A Conduit for Peace

K. David Hambright, F. Jamil Ragep, and Joseph Ginat

The fall of 2001 will long be remembered as a time of deep turmoil in the United States, the Middle East, and the world at large. In the midst of that turmoil, and in an especially sensitive political context, the Center for Peace Studies (CPS), a division of the International Programs Center at the University of Oklahoma, sponsored a conference on "Water in the Jordan Valley" bringing together natural and social scientific water experts from Israel, Palestine, Jordan, and the United States. Through the discussions that emerged out of this conference, the CPS, a consortium including Bethlehem University in Palestine, the University of Haifa in Israel, the Horizon Center in Jordan, and the University of Oklahoma, continued its role as a non-partisan facilitator, allowing representatives from countries with political disagreements to discuss in a neutral setting their differences on key issues. Previous conferences sponsored by the CPS dealt with such issues as Palestinian refugees (1999), the status of Jerusalem (2000), and the Middle East peace process (2000).

This conference was held on the campus of The University of Oklahoma, in Norman, Oklahoma, on November 13 and 14, 2001. Conference speakers included noted scholars from a variety of disciplines including engineering, agronomy, biology, economics, history, geography, and political science. The conference was organized into sessions with the following topics: (1) Water and Conflict: Historical and Cross-Cultural Perspectives; (2) Water and Peace in the Middle East: Prospects for Future Cooperation; (3) The Political Economy of Water in the Jordan Valley; (4) Water Shortages in the Jordan Valley: Proposed Technical Solutions; and (5) Water, Agriculture, and Environmental Sustainability. There was also an after-dinner address by Professor Franklin M. Fisher of MIT.

Among the issues raised by the speakers and that elicited considerable discussion were: (1) the viability of desalination, in particular its costs and in which cases it might provide a substantial share of regional needs; (2) the past, present, and future role of water in Middle East conflicts; (3) the possibility of regional solutions to water scarcity requiring cooperation among states; (4) the long-term prospects of the various aquifers and other freshwater sources; (5) environmental deterioration of water resources; (6) breakthroughs and developments increasing regional agricultural productivity, depending less on high-quality waters while turning to lower quality resources, such as recycled and brackish waters; and (7) alternatives to current water-usage patterns, particularly with regard to agriculture and the possibility of redirecting water to tourism and other sectors. From political and social realms to theoretical and scientific spheres, the purview of the conference embraced a wide and influential scope of timely issues.

The conference discussions and subsequent chapters in this book offer few definitive conclusions or facile solutions to increasing water supplies or distributing current water supplies equitably; yet there is a broad consensus that regional solutions to maximizing existing water resources must be pursued even as desalination becomes more viable both from technical and economic standpoints. The continuing deterioration of existing water supplies, both in terms of quantity and quality, mandate that any solution must be achieved within a political and social framework of peace, enlightened economic policies, and the sensible application of technical solutions that take due account of environmental concerns.

Among the most notable achievements of the conference, the gathering of participants not only from across political divides but also from across disciplinary boundaries represents a significant success. Several individuals remarked that they had rarely attended a conference in which they could debate with scholars from outside their professional fields. Thus agronomists were confronted with economists, for instance, who questioned the viability of agricultural subsidies; engineers listened to the complaints of biologists and environmentalists about overuse of resources and the impracticality of certain technical solutions; historians argued with economists and policy professionals about water as a cause of war; and scientists put forth their dreams of regional cooperation in research even as social scientists reminded everyone of the deteriorating political situation. This cross-fertilization also occurred cross-culturally, as several speakers highlighted comparative situations from other parts of the world (e.g. the western United States, the Mexican–US border region, and sub-Saharan Africa). All in all, there were many healthy exchanges fostered in an atmosphere of mutual respect that many participants hoped they could continue to develop in the future.

At some point in his or her life, nearly everyone has heard or read: "Water is the source of life." Those of us fortunate enough to have never experienced a prolonged period of water scarcity may agree with the obvious nature of that simple phrase, but nevertheless can find it difficult to fully appreciate its meaning. However, "Water is the source of life" wields considerable significance in many areas of the globe where drought and familiarity with the lack of potable water are common aspects of daily life. The Middle East, and particularly the Jordan River Valley, is one such region.

Many people and countries around the globe must cope with this most crucial problem of the lack of water. In this sense, the Middle East is not unique. However, when political disagreements between states combine with issues of fair and equitable access to common water resources, the problems associated with water scarcity become seemingly intractable. In this case, the Middle East is indeed unique.

In order to appreciate fully the complexity of these wide ranging issues, it is important to keep in mind the geographical and historical aspects that frame the discussion. Several regional and national maps, which readers will find helpful for general orientation, are: "The Mountain Aquifer" (figure 6.1, p. 74); "The Principal Production Water Wells in the Southern West Bank" (figure 7.1, p. 87); "Location Map Showing Salt Bodies" (figure 8.1, p. 102); "A Map of Israel and the National Water Carrier Network" (figure 10.1, p. 128); "West Bank Aquifers" (figure 11.1, p. 152); and "The Current Water Distribution System of the Jordan River" (figure 12.1, p. 165). The Jordan River receives water from three tributaries: the Banias River on the Golan Heights; the Hazbani River in southern Lebanon; and the Dan River in northern Israel. The Jordan flows into the Sea of Galilee in Israel and continues southward to the Dead Sea, forming the border between Israel and Jordan, as well as that between the West Bank and Jordan. Just south of the Jordan outlet from the Sea of Galilee, the Yarmuk River (on the border between Jordan and Syria) joins the Jordan River. Thus the waters of the Jordan River are shared by five countries, most of which are not at peace with Israel. In his chapter, "The Jordan Valley's Water: A Source of Conflict or a Basis for Peace," **Moshe Ma'oz** presents an overview of conflicts over water between Israel and its neighbors during the latter half of the twentieth century, arguing that without agreements over water allocation there can be no peace in the region. **Rateb Amro**, in "Historical Political Conflict of Jordan River Water Resources," examines details of the water quantities in the Jordan River basin and the major proposals and agreements that have been advanced in an effort to resolve what he believes has been, and continues to be, at the heart of the Israeli–Arab conflict – fair and reasonable entitlements to fresh water resources.

With respect to water resources, most proposals, understandings, and agreements between the various parties are guided in principle by the Johnston Plan of 1955 (also known as the Unified Plan for the

Development of the Jordan Valley), which specified water shares for the riparian parties, demarcated storage reservoirs, and called for international supervision of the implementation. "Compliance with and Violations of the Unified/Johnston Plan for the Jordan Valley," by **Munther J. Haddadin**, provides a detailed examination of the historical background of the Johnston Plan, the underlying negotiations between Ambassador Johnston and the riparian countries, and subsequent developments to the present day, including first-hand knowledge of negotiations between Jordan and Israel, and between Jordan and Syria. Haddadin also details several examples of compliance with and violations of the Johnston Plan by Jordan, Israel, and Syria.

Israel is not the only country in the world to receive a large proportion of its water from outside its boundaries: Egypt is downstream of Sudan on the Nile River; Iraq and Syria are dependent on Turkey for maintaining the flow of the Tigris and Euphrates Rivers; ten European countries draw water from the Danube River; India and Bangladesh share the water of the Ganges River.[1] In his chapter, "Water Resources Scarcity in West Africa: The Imperatives of Regional Cooperation," **Aondover Tarhule** details how access to and control of water resources in semi-arid West Africa have led to conflicts similar to those common in the Middle East. In his view, the study of how water scarcity induces or propagates conflict in West Africa may help mitigate or even prevent future conflicts in other regions of the world.

Of course, not all fresh water resources are at the Earth's surface. Roughly 25% of the world's fresh water resources is located in underground aquifers.[2] Though not obvious from the aforementioned conflicts over Jordan River resources, joint utilization and management of water derived from rivers may prove finally to be easier than the use of water derived from an aquifer. Where rivers actually flow, it is possible to determine riparian rights for the distribution of water. Yet the demarcation of boundaries and the determination of the amount of water present, as well as the potential amount available for extraction, represent much more complex issues for underground aquifers. One of the most challenging problems today lies in fact in the demarcation of and authority over the Mountain Aquifer shared by the West Bank and Israel. Both Israelis and Palestinians have long realized that the aquifer must have joint management, and this awareness resulted in the establishment (through the Oslo B Agreement) of the Joint Water Committee (JWC) and its arm of implementation, the Joint Supervision Enforcement Teams (JSETs). Basically, the JWC has de-facto veto power over any water resource development in the West Bank, as it must authorize all development initiated by any party. However, the effectiveness of the JWC and JSETs was compromised by the need for agreement on every action by representatives of two bodies (the Israeli Water Commissioner and the Palestinian Water Authority), which meet only occasionally, as well as the discrepancy of power between the two states.[3] On January 31, 2001, the JWC convened at the Erez Crossing (near Gaza) to sign an agreement stating that it was important to keep the water infrastructure out of the cycle of violence. **Eran Feitelson** re-evaluates the concept of joint management after several years of intifada. Given the paucity of experience of the global community in managing trans-boundary aquifers, he asks: "Is Joint Management of Israeli–Palestinian Aquifers Still Viable?" Feitelson discusses the current trend of separate management, which he claims represents a worst-case scenario with respect to resource management, and suggests that in order for Israelis and Palestinians to reach a form of cooperative agreement governing joint management of shared aquifers, trust must be re-established between the two parties.

Today's climate originating with the intifada, military incursions, and political posturing makes cooperation between Israeli and Palestinian water resource managers difficult, at best. However, the JWC remains in place, and there is a common level of respect for water-related infrastructure and personnel. For example, despite the state of belligerence in the West Bank, authorities have stated in Palestinian newspapers that water infrastructure workers from the Israeli National Water Company, Mekorot, are not to be harmed, nor are the Israeli vehicles connected to water supply to be disturbed.[4] Nevertheless, **David J. Scarpa** details the difficulties Palestinians continue to face in implementing official policy and resolutions when extracting water from the Mountain aquifer in "The Southern West Bank Aquifer: Exploitation and Sustainability."

Management of both surface and ground water resources is not simply a problem of resource quantity, but of its quality as well. In "Groundwater Salinization in the Jordan Valley – Quo Vadis?," **Akiva Flexer** and colleagues set forth the problem of groundwater salination arising from over-exploitation of the multi-aquifer system of the Jordan Valley. A similar challenge of declining water quality exists in surface waters as well. **K. David Hambright** and **Tamar Zohary**, in "Lake Kinneret and Water Supply in Israel: Ecological Limits to Operational Supply," offer a description of the obstacles facing the largest surface water reservoir in the region – Lake Kinneret (also known as the Sea of Galilee and Lake Tiberias). The authors show how the combination of exploitation and drought has reduced the water quality in the lake, thereby compromising the use of the water for domestic consumption.

Another major problem regarding water in the Jordan Valley and neighboring areas is identifying exactly how much water is available, when it is available, and what its major uses are. **Yoav Kislev**, in "The Water Economy of Israel," provides an excellent overview of Israel's current water resources and the levels of exploitation and consumption. **Alfred Abed Rabbo**, in "Current Water Provision and Allocation in Palestine," presents an analogous examination of Palestinian water resources and use. "The Peace Process and Water Supply in Jordan: Inter- and Trans-Boundary Border Projects," by **Mohammed Abudayyeh Matouq**, provides a detailed look at Jordan's water supply before and after the signing of the peace treaty between Jordan and Israel, focusing on the difficulties of meeting drinking water supply needs in Jordan's largest city, Amman, which holds more than 35% of the total population. **Mohammed Issa Taha Ali** analyzes the dynamics of water consumption in Jordan and suggests reforms in water pricing, distribution, and supply for improving the economic sustainability of water consumption in Jordan in his contribution, "An Economic Approach for Making the Most of Jordan's Water." A provocative argument is presented by **Franklin M. Fisher** in "Water: Casus Belli or Source of Cooperation?," a study in which the author suggests that water would represent a win–win scenario for all involved if only the parties were willing to distill down to economic valuation all arguments over water. **Onn Winckler**, in "Water, Demography, and Future Economic Development in the Triangle: Jordan, Israel, and the Palestinian Territories," pursues this line of reasoning, postulating that if the Jordan Valley countries phased out agriculture as a major component of their economies and developed instead a tourism-based economy, there would no longer be such substantial conflict associated with regional water access and allocation. An alternative to the purgation of agriculture is introduced by **Zvi Karchi** in "High Income Innovative Crops and Optimal Fertigation System: The Solution for High Farm Income under Water Shortage in the Jordan Valley." He demonstrates that agriculture need not represent a wasteful use of water and indeed that farmers can benefit economically from water-saving irrigation and fertilization technologies even when water is scarce. **Daniel J. Cantliffe**, in "Protected Agriculture: A Regional Solution for Water Scarcity and Production of High-Value Crops in the Jordan Valley," reveals how surprisingly simple it can be for greenhouses to increase water-use and land-use efficiency as well as overall farm production, in some cases as high as ten-fold.

Four years have passed since the November 2001 conference of Middle East regional water experts at the University of Oklahoma. Fortunately the Center for Peace Studies, with the financial assistance of the Citizens Exchange Program of the US Department of State, has actively built on the momentum sparked at the conference. During the past three years, working groups of water experts, community and political leaders, and students (both from the Middle East and the US) were established. Because of the political realities, two sets, a northern and a southern tier, worked in parallel: one group consisted of Syrians, Lebanese, Jordanians, Turks, Iraqis, and Americans, while the other included Israelis, Palestinians, Jordanians, and Americans. Jordanians thus served as a kind of bridge between the groups, while the Americans acted as facilitators. As with the conference, the groups represented a variety of disciplines and focused on technical details. At the same time, they made use of the opportunity for trust-building and for establishing networks of communication both among experts and students. The following passage is from the report submitted after the first 2003 meeting of the Southern Tier Water Working Group:[5]

> Participants agreed that water is critical to peace in the Middle East because it is shared by all states in the region. Further, participants agreed that all states in the region, not just those in the Southern Tier, must cooperate to develop solutions to water shortages. Trust between all participants was identified as critical to regional cooperation. The Water Working Groups Project was praised as an important tool in building trust among experts in the field of water and in generating equitable solutions to the water crisis. Participants identified technological solutions, public education, and water conservation at the level of the individual and community as essential components of the solution to the regional water shortage. Participants maintained that states will have to consider desalination and trans-boundary water transfers in the future in order to provide adequate water resources for everyone in the region. Regional education and conservation were identified as crucial for alleviating the water shortage. The political situation of the Palestinian people has created a unique situation in regards to water resources. Some of the associated challenges (e.g., lack of water supply infrastructure) in Palestinian areas were discussed. Finally, the creation of water-efficient model villages was proposed as a method of developing, implementing, and demonstrating the technological, educational, and conservation ideas discussed at the meeting. It was suggested that model villages be located in each of the Southern Tier countries to serve as models for access and prudent use of water. Construction and development of the proposed model villages would be an entryway for Business, Governmental, and NGO Partnerships that would be managed and implemented on a regional, multi-lateral scale. Participants suggested that future working groups research possible funding and partner agencies as well as the political and technological realities of model villages.

And in the Northern Tier, an exciting initiative took shape that culminated in an agreement signed in May 2005 by which Iraqis, Syrians, Turks, and Americans agreed to work together to improve "the quality of life for people in all communities, including rural and urban areas" and to promote "harmony among countries and with nature in the Euphrates–Tigris region." This "Euphrates–Tigris Initiative for Cooperation" would have been unthinkable even two years ago.

We look forward to more of these meetings as an effective means of promoting exchange, trust, and cooperation, not only among technical experts dealing with water, but among the entire citizenry of all Middle

Eastern countries. It is our hope that the following chapters will provide a sound basis for optimism and a realistic sense that peace can indeed be achieved. Water is vital to life – in the Middle East, water is also vital to peace.

Notes

1   Yaron London, *Yediot Aharonot*, June 3, 2002.
2   T. V. Cech, *Principles of water resources: history, development, management, and policy* (New York: John Wiley & Sons, 2003), p. 446.
3   M. Haddad, E. Feitelson, S. Arlosoroff and T. Nasseredin, A Proposed Agenda for Joint Israeli–Palestinian Groundwater Management, in Eran Feitelson and Marwan Haddad (eds.) *Management of Shared Ground Water Resources: The Israeli–Palestinian Case with International Perspective* (IDRC Kluwer Academic, 2001), p. 500.
4   Personal communication to Prof. J. Ginat from Dr. Joseph Guttman, Mekorot Israel National Water Co. The same description was received from Khaled (pseudonym), a high official in the Palestinian Authority who asked that he not be identified.
5   M. Chumchal and K. D. Hambright, "Working groups to study water resources and usage in the Middle East, southern tier meeting 1: Israelis, Jordanians, and Palestinians," Report submitted to the University of Oklahoma International Programs Center, Center for Peace Studies (2003).

Part I

Historical and Political Perspectives

CHAPTER 2

The Jordan Valley's Water: A Source of Conflict or a Basis for Peace

MOSHE MA'OZ

Water Plans: From Lowdermilk to Johnston

In a book published in 1944, Dr. W. C. Lowdermilk, an American soil conservationist, outlined his visionary plan for a "Jordan Valley Authority" which would create a single unit for development on the lines of the Tennessee Valley Authority.[1] In his introduction to a book on the same plan in 1948 by the American irrigation engineer James B. Hays, Lowdermilk wrote: "There are few places in the world where mankind has a more favorable opportunity to adopt a constructive approach towards the problem of the common man, removing the basic causes of conflict and war by the creation of abundance for all. We can, through this approach, make the Middle East a blessed example rather than a breeding ground for strife."[2]

To be sure, the implementation of this grand Jordan Valley plan, like similar projects, required close cooperation among Syria, Jordan, Lebanon, and Israel. However, in May 1948 the first Arab–Israeli war erupted and, despite the armistice agreement in 1949, these Arab nations continued to be in a state of war with Israel. Accordingly, the water plans of these enemy states were prepared separately with little or no consideration of the water interests of the "hostile" neighbor.[3] Not only did these Arab states and Israel object to, and complain about, the opposite water schemes and the relevant works undertaken, they also rejected a UN-sponsored regional water plan for the Jordan Valley, prepared by Chester T. Main Inc. (USA) under the supervision of the Tennessee Valley Authority.[4]

However, this latter plan (September 1953), which failed to satisfy the water interests of the relevant states, served as a blueprint for an official

American plan – the Eric Johnston Jordan Valley project – also aimed at settling the water conflicts between Israel and its Arab neighbors. Initially (from late 1953) this plan encountered harsh criticism by both Arabs and Israelis, each side alleging that the plan favored the other. Israel's Prime Minister Ben-Gurion depicted the plan as an "American mandate" and pro-Arab, while Arab leaders rejected it for "fear that the acceptance of the plan would imply acquiescence in the existence of Israel."[5] Subsequently, however, both sides agreed to examine the Johnston plan and present counter-proposals.

The Arab League, led by Egypt, was now prepared to include Israel indirectly in the project, but to allocate it very little water, while refusing to turn Lake Tiberias into the regional water reservoir. Israel's proposal was similarly not generous to Arab water needs, and also requested that the Litani River (in Lebanon – neither an international river nor a tributary of the Jordan River) should be included in the Jordan Valley project.[6] Israel, which had already started to design its Jordan–Negev project, also objected to international supervision of the Jordan Valley plan and continued to suspect Arab ill intentions. But eventually, in late 1955, Israel approved a revised Johnston plan that excluded the Litani River.

The Arab states first showed some flexibility, naming the plan the "Arab plan," but subsequently they split into two camps. Egypt and Jordan accepted the revised Johnston plan while Syria and Lebanon rejected it. General Burns, chairman of the Syrian–Israeli Mixed Armistice Commission, wrote later:

> The Johnston negotiations seemingly close to success were stalled by the obduracy of the Syrian politicians. They simply would not agree to anything that would benefit Israel, even if the Arab states would thereby achieve greater benefits. Syria also opposed anything implying recognition of Israel's right to exist.[7]

Accordingly, the prime ministers of Syria, Lebanon, and Jordan convened in Damascus in early October 1955 and decided to reject the Johnston plan and prepare a new Arab plan. But on 12 October, under Egypt's inducement, the Arab League's political committee reversed the Damascus resolution and decided to further reexamine the plan and reach a conclusion that would be "compatible with Arab interests." Around the same time, Ambassador Eric Johnston told Israel's Minister of Foreign Affairs, Moshe Sharett, that Gamal Abd al-Nasir (Nasser), Egypt's president, had asked that Israel should delay its works to divert the Jordan River for two or three months in order to have Syria join an agreement regarding the Jordan Valley plan. Sharett agreed to wait even longer, until March 1956, stipulating that if no agreement were reached, Israel would unilaterally start work to divert the Jordan River.[8]

The Israeli–Jordanian Model

As it happened, no Arab–Israeli agreement was reached regarding the Johnston plan; however, there was a tacit understanding, or unwritten agreement, apparently under American auspices, between Israel and Jordan, namely, that each country would utilize the Jordan Valley's water, not exceeding its share according to the revised Johnston plan.[9] This tacit agreement was secretly kept by the two sides for several decades until it was replaced officially by a water agreement within the Israeli–Jordan peace treaty of 1994.

This Jordanian–Israeli model of *de facto* cooperation or coordination, despite the *de jure* state of war between these nations, rested on four legs: (1) the vital dependence of both nations on the Jordan Valley's water and the urgent need to utilize it for economic development and unusual demographic growth, i.e., Palestinian refugees in Jordan and Jewish refugees and immigrants in Israel; (2) the role of the United States in initiating and financing the projects, as well as mediating between the parties; (3) the pragmatic policy of Egypt that rendered diplomatic backing to Jordan in the Arab League despite popular Arab criticism. Egypt's leader Nasser was reluctant to use the dispute over the Jordan Valley's water as a *casus belli*, since Egypt's army was not prepared for war with Israel; and (4) common strategic interests and periodic cooperation between Israel and Jordan, notably *vis-à-vis* Palestinian radicalism, which started before the 1948 war, continued in the 1950 attempt to reach a peace agreement, and culminated in the September 1970 event ("Black September") when Israel helped Jordan to survive a Palestinian rebellion and a Syrian invasion.

Lebanon and Syria versus Israel

As indicated above, Lebanon and Syria, unlike Jordan, strongly objected to the Johnston plan as well as to the diversion of the Jordan River by Israel. Each of these states threatened to divert the Jordan River's tributaries – the Hazbani in Lebanon and the Banias in Syria – in order to prevent Israel from using their waters. Even though these rivers-tributaries were of little significance in the water systems of these countries, their diversion was very costly and could also damage Jordan's water interests.

However, beyond the vocal threats and inflammatory rhetoric, the real agendas of Lebanon and Syria were totally different. With its largely Christian Maronite character, Lebanon was anxious to demonstrate by words, but not deeds, its pan-Arab anti-Israeli attitude. But, with its weak army, it had no intention to divert the Hazbani and thus provoke an Israeli military attack. Indeed, until recently the Hazbani River has not been a source of conflict between Israel and Lebanon, although Lebanese politicians have periodically accused Israel of attempting to divert or to steal the Litani River waters.[10]

By comparison, the water disputes between Syria and Israel reflected the mutual ideological, strategic, political, and psychological antagonism between the two states. These disputes constituted the major factor that led to the June 1967 Arab–Israeli war and have remained a main obstacle to a Syrian–Israeli peace.

The Syrian–Israeli Water Conflicts

Ideologically, Syria has been for many decades one of the most antagonistic Arab nations towards Zionism and Israel. This is owing to its claim to be the "beating heart of Arabism" – the pioneer of the Arab nationalist movement and the birthplace of the Ba'th pan-Arab movement which has been in power in Damascus since 1963. Ideologically, historically, and strategically, Damascus has for generations considered Palestine (including Israel, and also Jordan and Lebanon) as part(s) of "Greater Syria." But although Syria was a major initiator of the 1948 and 1967 (and 1973) wars with Israel, it was too weak and would not dare fight Israel single-handedly, but was bound to form military coalitions with Egypt and other Arab states. against Israel. Yet Syrian leaders were bold enough to threaten Israel with annihilation, to coordinate an economic boycott against it, and periodically to launch military attacks against Israeli positions, villages, and fishermen along the Jordan Valley and Lake Tiberias. During the mid-1960s, Syria also employed *Fatah* – Palestinian guerrilla fighters – to attack Israeli water installations and other civilian targets, aiming to provoke a "popular liberation war" and to drag Egypt into an all-Arab military offensive against Israel.

To be sure, Israel by no means remained passive *vis-à-vis* Syrian threats and actions. Considering it an arch enemy, Israel was not content with warning Damascus but periodically launched massive and painful attacks against Syrian military positions across Lake Tiberias and the Jordan Valley, as well as against the Banias River diversion site. Several such attacks were unprovoked, possibly aiming at generating an all-out war with Syria (and Egypt). Indeed, Israeli leaders, notably David Ben-Gurion and Moshe Dayan, were convinced that Syria (and Egypt) intended to destroy Israel, not to mention prevent it from undertaking water projects in the Jordan Valley that were destined to develop Israel's economy and absorb hundreds of thousands of Jewish refugees and immigrants. Consequently, Israel has been determined not to compromise important strategic assets in the Jordan Valley, even in return for peace with Syria, and to unilaterally design and execute its Jordan Valley water projects, regardless of Syrian interests.

Disputes over Lakes Hula and Tiberias

During the 1948 war, Syria occupied several strategic areas along the Jordan Valley that had been allocated to Israel in the UN Partition Resolution No. 181 of 29 November 1947. These areas were situated at the southeastern and northeastern banks of Lake Tiberias, the Jordan Valley south and east of the Hula Lake, and at the slopes of the Banias River. And although these zones were demilitarized in the 1949 Syrian–Israeli Armistice Agreement, Syria dominated large parts of them, owing to its topographic advantage; and in 1951 it took by force the al-Himma area, south of Lake Tiberias, near the junction of the Yarmuk and Jordan rivers. No wonder then that Prime Minister Ben-Gurion rejected peace offers by Syrian dictators, Husni Za'im in 1949 and Adib Shishakli in 1952, in return for Israeli concessions in Lakes Tiberias and Hula as well as al-Himma.[11]

Already by early 1951, Israel had embarked upon a large development project – the drainage of Lake Hula, fed by the Jordan River, in order to reclaim 15,000 acres for cultivation, eradicate malaria from the area, and increase water supply. But apart from the economic benefits, the project also aimed at strengthening Israel's position in the adjacent demilitarized zone (DZ), where ground works started. As Israel's chief of staff Yigael Yadin stated:

> When we decided to drain the Hula and began to work, we had to start in the demilitarized zone, of all places . . . If the project weren't necessary for its own sake, we would have done it for political reasons.[12]

Syria obviously objected to the project on the grounds that it was executed in Arab-owned land in the DZ, and that it would obliterate a natural ground obstacle and grant Israel a military advantage. The UN Mixed Armistice Commission (MAC) rejected Syria's latter claim but ordered Israel to cease the work within the DZ.

However, Israel continued working in the DZ amidst fire exchanges with Syrian proxies and forcibly evacuated the inhabitants of two Arab villages in the DZ. Syria reacted by killing Israeli soldiers dispatched to al-Himma village, and occupied the village. In retaliation to Israeli air strikes on one of its border positions, Syria occupied Tal al-Mutilla, a strategic hill north of Lake Tiberias, one mile within Israel's side of the border. Israel subsequently recaptured the hill with heavy losses, and later completed the Hula project, in accordance with the Security Council resolution, but without working on Arab land in the DZ.

In comparison with the Hula Lake conflict, which did not last long, the Lake Tiberias disputes generated greater repercussions and continued until the June 1967 war. For Israel, Lake Tiberias has been a centerpiece of its national ethos and the major reservoir of its water. For Syria, which topographically dominated the lake from the Golan Heights, it has been a strategic target *vis-à-vis* Israel, as well as a source of water and fish for several local villages. Syria, which *de facto* controlled the eastern shore of the lake, including the 10-meter strip in the northeastern tip which was formally under Israeli sovereignty, prevented by force Israeli fishing in that part of the lake. Israeli police boats did the same to Syrian fishermen at that part of the lake and occasionally were engaged in shooting incidents with Syrian military positions. Considering the struggle with Syria as a zero-sum game, the Israeli army, under Prime Minister Ben-Gurion's orders, undertook by early 1954 "to react in an aggressive manner in each case affecting Israeli fishing in the Kinneret" [i.e. Lake Tiberias]; this was in lieu of permitting Syrian fishing at that spot, thus creating a balance of interests as Foreign Minister Sharett may have advocated.[13]

Furthermore, it would seem that by late 1955 Ben-Gurion had attempted to use these fishing disputes in order to provoke a major battle with Syria, defeat its troops, which had just received new arms from the Soviet Bloc, as well as test its October 1955 military pact with Egypt (directed against Israel). Indeed on 10 December 1955 Israeli police boats patrolled near Syrian positions on the northeastern shore of the lake, provoking them to open heavy fire.[14] And on the night of 11 December, Israeli troops (led by Major Ariel Sharon and supported by aircraft and artillery) attacked several Syrian positions at Kursi and Butayha, killing 37 Syrian soldiers and 12 civilians, while taking 30 prisoners. However, this massive raid, which was severely criticized by Foreign Minister Sharett and other Israeli officials and drew international condemnations, failed to deter Syria, even aggravating its conflict with Israel. Thus, following this raid, Syria allocated more resources to its military build-up, acquired new arms from the Soviet Union, and strengthened its military and political relations with Egypt, culminating in the February 1958 union – the United Arab Republic (UAR) – between these two states. During the period of this union, until September 1961, there occurred hundreds of shooting incidents between Syrian and Israeli troops over fishing rights in the lake, over Israeli land developments in the southern DZ, and particularly over the Israeli Jordan–Negev water project.

Conflict over the Jordan–Negev Water Project

Already in 1953 Israel contemplated diverting the water of the Upper Jordan River to the Negev desert region, which comprised 60% of Israel's uninhabited lands. The first phase of the project included digging a canal from Bnot Ya'acov bridge to Lake Tiberias and building a power station that would use the water flow in the canal to produce electricity. Drawing a lesson from the Hula project, Israel stated that it would not use Arab-owned land in the DZ, where part of the new canal was designed to pass. Syria again objected to the project mainly on the grounds that the military balance in the region would change in Israel's favor, owing to topographical alterations. Syria also threatened to use military force and divert the Jordan tributaries should Israel carry out the project. The joint Syrian–Israeli MAC backed Syrian arguments and requested Israel to stop working on the project. After Israel ignored this request, the US threatened Israel with economic sanctions and launched its own Johnston Jordan Valley plan for the allocation of the Jordan and Yarmuk River waters among Israel, Jordan, Syria, and Lebanon.

Israel accepted the revised Johnston plan, as indicated above, and stopped its Bnot Ya'acov project. But since Syria eventually rejected the Johnston plan, Israel started work in early 1958 on the Jordan–Negev diversion project at a new site, namely, the northwestern corner of Lake Tiberias which was under full Israeli sovereignty. Syria reacted by sharp verbal attacks on Israel, by intense diplomatic activities in the Arab League, and by exercising pressure on Nasser to stop the diversion project by force, but all to no avail. Simultaneously, Syrian troops continued to fire at Israeli police boats on the lake and also shelled Israeli villages around the lake. Israel again retaliated in a sharp way, launching two massive attacks on Syrian positions at Tawafiq (southern DZ) in late January 1960 and at Nuqaib (northeast of the lake) in mid-March 1962.[15]

In March 1963 a new and more critical phase started in the Syrian–Israeli conflict over the Jordan–Negev diversion project. Ba'thist nationalist-radical officers seized power in Damascus and declared that this project "must be prevented by force" and that Israel must be liquidated.[16] In Israel a new government was formed in 1963, headed by Levi Eshkol, a moderate, hesitant leader, but several of his ministers and army officers urged him to adopt an activist-militant policy towards Syrian belligerency.

To be sure, the new Syrian Ba'thist regime went beyond hostile rhetoric and initiated many shooting attacks against Israeli troops, who reacted in kind. Syria also trained Palestinian *Fatah* guerrillas and dispatched them to damage a section of the Jordan–Negev water carrier on 1 January 1965, with no success. Simultaneously, Syria started work to divert the Banias River, but Israel's air force bombed and destroyed the diversion equipment in early 1965. By then, more realistic Syrian Ba'th leaders acknowledged that their army was incapable of fighting Israel single-handedly, and this was also the case of all Arab armies united. As a senior Ba'th leader admitted:

[We] evaluated that the Syrian army, despite its good arms, experience and
courage, was not in a position to hold out more than a few hours against any
Israeli attack . . . The Arab leaders attending several meetings . . . [regarding]
the Palestinian problem . . . [and] the diversion of River Jordan . . . had to
face the bitter truth. They were leaders of a nation of a hundred million Arabs
who were hardly armed, facing a country of a million and a half which is fully
armed.[17]

Nevertheless, the Syrian regime endeavored time and again, since 1963,
to drag Egypt into a war with Israel. But Nasser was not prepared to fight
Israel over the Jordan River diversion, since 40,000 of his troops had been
engaged in a futile war in Yemen. He managed to outwit the Syrian leaders
by initiating resolutions at Arab Summits to refrain from war against Israel
and only engage in diverting the Jordan tributaries into Arab lands.
Subsequently, however, in May 1965 Nasser "acknowledged that the Arab
diversion plan could not be carried out and that the Arabs could not go to
war in the foreseeable future."[18]

Only in late 1966 did Nasser gradually change his position, when the
Syrian–Israeli water conflict seemed to him to escalate into an all-out
Israeli offensive against Syria. The new Ba'th (neo-Ba'th) regime in
Damascus, led by Salah Jadid and Hafiz Asad since February 1966, was
more militant than the previous regime. It was determined to generate a
popular liberation war against Israel, pioneered by the Palestinians and
turning into a conventional war by Arab armies, led by Syria and Egypt.

Thus, in addition to initiating numerous border incidents, Syria organized or directed dozens of *Fatah* guerrilla/terrorist operations inside
Israel, mostly via Jordan, while stepping up strong verbal attacks against
Israel. These anti-Israeli words and actions were certainly provocative,
even according to those foreign analysts who supported the Syrian cause:
"By carrying terrorism into Israeli territory . . . it [Syria] promotes guerrilla
warfare and yet cannot protect itself against [Israeli] reprisals" and "the
increasing evidence of Syria's verbal aggressiveness more and more
assumed the character of a calculated provocation, as the neo-Ba'th came
to believe that the time had come to engage Israel in decisive battle. This
provocative policy gave the Israeli activist clan precisely the pretext it
needed to go to war to fulfill its territorial ambitions."[19]

Indeed, Israeli activist leaders – Moshe Dayan, Shimon Peres, and
Yigal Allon – pressed Prime Minister Eshkol to inflict a massive military
blow on Syria. But the pragmatic Eshkol opted for limited operations
against Syria for the "useful purpose of demonstrating Israeli capability
of interdicting the diversion works by measures short of full-scale war."[20]
He did not want to antagonize the US government that stated "its support for Israel's water project within the quantities of the unified
[Johnston] plan, and its opposition to Arab diversions contrary to this

plan . . . [But] the US simply could not accept Israel's military intervention against the Arab projects."[21]

Nevertheless, Israel's chief of staff General Yitzhak Rabin issued in May 1966 a threatening statement towards Damascus: "Israel's reaction against Syrian activities must be directed against those who carry out sabotage and against rulers who support these acts . . . Hence the problem with Syria is basically one of a clash with the rulers." Or, according to the Syrian version: "The attacks that Israel is being forced to make in reprisal for the sabotage raids . . . are thus aimed at the regime in Syria."[22]

Whether or not Rabin intended to cause the collapse of the neo-Ba'th regime, from July 1966 Israel escalated its reaction to Syrian attacks including the use of air power against the Banias diversion site and other Syrian military positions. In dog fights that subsequently occurred, the Israeli air force shot down several Syrian MiG planes and, on 7 April 1967, Israeli war planes flew over Damascus. Earlier, on 13 November 1966, Israeli troops massively attacked Samu' village in the Jordanian-held West Bank in retaliation for anti-Israeli guerrilla/terrorist operations emanating from the West Bank. This raid provoked fierce verbal attacks on the Jordanian regime by Palestinians, Syrians, and Egyptians, and it also induced this regime to join later (on 30 May 1967) the newly-signed Egyptian–Syrian Defense Pact (of 7 November 1966). On 14 May 1967, Egypt, invoking its fresh military pact with Syria, moved its troops into Sinai in reaction to "huge Israeli troop concentrations on the Syrian borders. Their intention is to intervene on Syrian territory in order to overthrow the Arab liberated regime in Syria."[23]

These Egyptian assertions were derived from Soviet reports regarding alleged Israeli troop concentrations, as well as on a new verbal threat against Damascus by General Rabin on May 14, 1967: "In Syria the problem is different because there it is the authorities who send out the saboteurs. Therefore, the aim of action against Syria is different from what it ought to be against Jordan and Lebanon."[24]

Not only did Egypt dispatch its troops into Sinai, but it also evicted the United Nations Emergency Force and closed the Tiran Straits for ships sailing to Israel's port of Eilat. Coupled with Nasser's highly belligerent speeches against Israel, the chain of events was set for the eruption of the June 1967 war.[25] Israel, which considered the Egyptian moves as a *casus belli*, started the war on 5 June 1967 and in six days conquered Sinai, the Golan Heights, Gaza Strip, and the West Bank. On June 15, 1967, the Israeli cabinet unanimously adopted the following resolution:

> Israel stands for the conclusion of a peace treaty with Syria [and Egypt] on the basis of the international boundary . . . the conditions for a peace treaty are: (1) a total demilitarization of the Syrian [Golan] Heights . . . (2) an absolute guarantee for free water flow from the River Jordan sources into

Israel either by an alteration in the northern boundary or by an agreement between the two countries.[26]

As we know, no agreement up to now has been reached between Syria and Israel regarding the flow of the Banias River from the Golan Heights into the Jordan River. Israel has controlled the Golan since June 1967, except for several days during the October 1973 war, and considers the Banias water as one of its important strategic assets.[27] Syria, for its part, refused for many years to negotiate with Israel on the status of the Banias water, like other Golan issues, before Israel would make a commitment to withdraw to the 4 June 1967 line. This would mean *inter alia* allowing Syria to again control the northeastern shore of Lake Tiberias – a non-starter for Israel.

Water and Peace

Even at the Madrid peace conference (October 1991), when Syria agreed for the first time to negotiate directly with Israel without pre-conditions, it demonstrated its hard line regarding the water issue. Reacting to Israel's attempt to attend a meeting on regional water problems, Syria's vice president 'Abd al-Halim Khaddam stated: "Israel has no right to a single drop of water in the region. It is absolutely unacceptable for Israel to be a party to any arrangement on water or any other issues in the region."[28]

Nevertheless, in December 1995, during an advanced phase in the peace negotiations at Wye Plantation, Maryland, Syrian and Israeli officials reached a "general understanding that water needs of both sides should be secured, regarding quantities and quality. Concerning Israel, this meant a guarantee for the continuation of its current water supply through a mechanism to be decided upon in future negotiations."[29]

Unfortunately, this phase of negotiations did not produce a peace agreement. This was the case also regarding the most recent phase of the negotiations in March 2000, under the auspices of US President Bill Clinton. The major issue that remained unresolved and halted the negotiations was related to Lake Tiberias. Syria insisted that Israel should withdraw to the 4 June, 1967 line, also on the northeastern bank of the lake (10 meters away), and render Syria free access to the lake for fishing and swimming, although not for pumping water. Rejecting this demand, Israel insisted on controlling a 100-meter-wide strip of this part of the shoreline. Both Syria and Israel rejected a compromise proposal, that the disputed strip of shore should be made a tourist region for both Syrians and Israelis, under joint sovereignty.

In conclusion, peace between Israel and Syria cannot be achieved without settling the water problems between them, namely the Banias

River/tributary and the northeastern shore of Lake Tiberias. Regarding the latter problem, it would seem that the above compromise proposal may serve as a basis for a mutually agreed settlement. The Banias River issue can be settled according to the Israeli–Jordanian model, stipulated in the 1994 peace treaty.[30]

Notes

1   W. C. Lowdermilk, *Palestine, Land of Promise* (New York: Harper, 1944).
2   M. G. Ionides, The Disputed Waters of Jordan, *The Middle East Journal* 7 (2) (Spring 1953): 156.
3   Frank Meissner, Prospectives for Artificial Rain Enhancement in the Jordan Valley Development, *The Middle East Journal* 7 (4) (Autumn 1953): 489.
4   Yoram Nimrod, *Angry Waters: Controversy Over the Jordan River* (Givat Haviva, 1966), pp. 37–39 (Hebrew).
5   See, respectively, *Ben-Gurion Diary* (Sde Boker), June 26 and August 25, 1954; Dulles to Eisenhower, confidential, May 7, 1954, Declassified Documents (USA).
6   Nimrod, *Angry Waters*, pp. 44–49.
7   Edson Burns, *Between Arab and Israeli* (London, 1962), p. 113.
8   Moshe Sharett, *Personal Diary* (Tel Aviv, 1978), V, 1357 (Hebrew); see also PRO London FO 371/121825, February 2, 1956.
9   *New York Times*, October 19, 1958; see also G. E. Gruen, *Water and Politics in the Middle East* (New York, 1964), p. 6, quoted in Nimrod, p. 82.
10   See, for example, Tabitha Petran, *The Struggle over Lebanon* (New York, 1987), p. 305. Cf. Reuven Erlich, *The Lebanon Tangle* (Tel Aviv, 2000), p. 557 (Hebrew).
11   Moshe Ma'oz, *Syria and Israel: From War to Peacemaking* (Oxford, 1995), pp. 20–26.
12   Quoted in Aryeh Shalev, *The Israel–Syria Armistice Regime, 1949–1955* (Tel Aviv University, 1993), pp. 51–52.
13   Aryeh Shalev, *Cooperation under the Shadow of Conflict* (Tel Aviv, 1989), p. 298 (Hebrew). Cf. Ma'oz, *Syria and Israel*, p. 50.
14   Burns, *Between Arab and Israeli*, pp. 110, 118; Damascus Radio, October 22, 1955.
15   See respectively FO 371/151200, February 1 & 18, 1960; *UN Year Book 1962*, pp. 135 ff.; S/5093 and S/5098, March 19 & 21, 1962.
16   Petran, *The Struggle over Lebanon*, p. 191; *Dirasat Ta'rikhiyya... li-Nidal Hizb al-Ba'th* (Damascus, 1972), p. 85.
17   Sami al-Jundi, *Al-Ba'th* (Beirut, 1969), pp. 164–66.
18   Petran, *The Struggle over Lebanon*, p. 193; cf. al-Jundi, *Al-Ba'th*, p. 165.
19   See, respectively, Patrick Seale, *The Observer Foreign News Service,* May 19, 1967; Petran, *The Struggle over Lebanon*, pp. 195–96.
20   US Department of State, May 25, 1965. Cf. *New York Times*, October 25, 1965.
21   Ma'oz, *Syria and Israel*, pp. 85, 87 and sources.
22   See, respectively, *Middle East Record* (1967), p. 161; *UN Security Council Official Records* (1967) S/7495, September 15, 1966.
23   *Middle East Record* (1967), p. 185. Cf. *New York Times*, May 26, 1967.

24   *Middle East Record* (1967), pp. 179, 186; *New York Times*, May 14, 1967.

25   For details, see Ma'oz, *Syria and Israel*, pp. 88ff.

26   M. Avidan, 19 June 1967: The Israeli Government Decides Hereby, *Davar* (Hebrew, Tel Aviv), June 2, 5 & 19, 1967.

27   General (Ret.) A. Kahalani, *The Negotiations Between Israel and Syria* (Bar Ilan University, 1993), pp. 9–14 (Hebrew); *New York Times*, July 17, 1991.

28   Foreign Broadcast Information Service (FBIS) (USA), October 10, 1991.

29   Uri Savir, *The Process* (Tel Aviv, 1998), p. 313 (Hebrew).

30   Lori Plotkin, *Jordan–Israel Peace: Taking Stock 1994–1997* (Washington DC, 1997), pp. 39–40.

CHAPTER 3

Historical Political Conflict of Jordan River Water Resources

RATEB AMRO

Contrary to a widely held view in the West, the most highly prized resource in the Middle East is not oil but water.[1] It is not only a key input to the many agricultural and industrial activities upon which development in the region depends, it is essential for life itself. And as populations and expectations of higher living standards rise, each drop of this scarce commodity will become even more vital to the peoples of the Middle East.

Unfortunately, most of the natural sources of this increasingly precious resource overlap the territories of competing and often openly hostile neighbors. For example, Israel, Jordan, Lebanon, Syria, and the West Bank are all riparians of the Jordan River basin; Iraq, Iran, Syria, and Turkey share the Tigris–Euphrates basin and nine Middle Eastern African states – Burundi, Egypt, Ethiopia, Kenya, Rwanda, Sudan, Tanzania, Uganda, and Zaire – depend to a greater or lesser extent upon the waters of the Nile River and its tributaries. Sharing common ground and surface-water was not a problem in years past when these renewable resources were more than sufficient to sustain the small populations and austere lifestyles of the region. However, the combination of expanding economic activities and extraordinary population growth has caused the domestic, agricultural, and industrial demand for water to escalate over the last half-century.

Consequently, states can no longer fully satisfy their needs from common resources without adversely affecting the quantity or quality (or both) of the supply available to others. In such circumstances, cooperation among users of these resources is imperative. Like water, though, cooperation is a scarce commodity in the Middle East. Many states and political communities are entangled in longstanding, bitter disputes with their neighbors that impede cooperative development and management of shared rivers and aquifers. Complicating the matter further is the absence of international agreements or conventions establishing each user's entitlement to these water resources. Consequently, development of these resources (e.g., the construction of dams, digging of canals, laying of pipelines, and drilling of wells) has often been done unilaterally based on the relative power of the users rather than on the principles of international law. As amply demonstrated in the past, this path can lead to open conflict.[2]

Since 1994, Arabs and Israelis are trying to break this cycle of conflict within the framework of the Middle East peace negotiations. Barring a collapse of the process, they will at some point have to confront the complex problem of water, including the definition of entitlement to shared fresh water resources. The challenge, however, will be to negotiate a fair and reasonable assignment of entitlements.[3]

The Jordan River Basin [4]

The headwaters of the Jordan River originate in southern Lebanon (the Hasbani River), northern Israel (the Dan River) and Syria (the Banias River) (table 2.1).[5] These three spring-fed rivers merge at a point six km south of Israel's northern border to form the upper Jordan River. The discharge of the upper Jordan River passes through the Huleh Valley and falls more than 200 m to Lake Tiberias (syn. Lake Kinneret, Sea of Galilee). This lake, lying at an elevation of approximately 210 m below sea level, is the principal natural storage reservoir in the basin.

Table 3.1   Contributions of streams to the Jordan River

Streams contributing to Jordan River	Average Annual Flow (Mm³)
Dan	245
Hasbani	138
Banias	121
Yarmourk	500
Intermittent Tributaries and Spring Waters	350
Total	1354

Lake Tiberias has a surface area of 170 km² and a volume of approximately 4,000 Mm³. Water enters the lake from the upper Jordan River as well as from seasonal streams or wadis. Local salty springs on the north western shore of the lake are diverted to the lower Jordan River so as to limit salinity levels in the lake. Ten kilometers south of Lake Tiberias, the flow of the Jordan River joins with that of its principal tributary, the Yarmouk River, and is fed by winter precipitation over the catchment area, with some spring discharge. The Yarmouk River, arising in Syria and Jordan, forms the international border between Syria and Jordan and at its lower reaches, between Jordan and Israel, until it merges with the Jordan River. South of the confluence with the Yarmouk River, the Jordan River flows through the Jordan Valley to the Dead Sea at 395 m below sea level. Along the way, it receives additional water from winter rainfall and perennial spring flow on the east and west banks of the river. However, due to salinity levels of several thousand parts per million, the water of the lower Jordan is unfit for domestic or agricultural consumption.

The catchment area of the Jordan River (including the Yarmouk River) is approximately 17,665 km². Its natural flow (i.e., in the absence of extractions) is estimated to be roughly 1,476 Mm³ at the entrance to the Dead Sea; this represents approximately 1.8% of the average annual flow of the Nile River at Aswan, Egypt. Despite its relatively small discharge, the waters of the Jordan River system represent important components in the water budgets of its riparians, particularly Israel, Jordan, and potentiality, the West Bank.[6]

Background [7]

Like the good earth itself, water in the Middle East is many things. It is the source of life and of riches, a fundamental economic base and asset and a symbol of sovereignty and a political weapon as well. Water, from 1947 to the present, lies at the very heart of the Israeli–Arab dilemma. Since the ill-fated United Nations (UN) efforts to solve that dilemma began, attempt after attempt to draw viable armistice lines or write peace agreements that would last have been blocked by the water question. While the need for rational international water-sharing schemes is glaringly apparent, such schemes become more difficult to attain. This is due to the demographic changes wrought by immigration, as well as by intransigence on both sides about whether and when Israel must yield up its 1967 territorial conquests. Added to this is the fact that Israel consumes at least five times as much water per capita as each of its less industrialized and less intensively farmed neighbors.

Even before World War II ended in 1945 and Israel's subsequent creation in 1948, Zionist leaders, determined to create a viable, independent Jewish state in Palestine, realized that the key to their success lay in exploiting the Jordan Valley's still untapped water, electric power, and agricultural riches. In 1944, an American water engineer, Walter Clay Lowdermilk, after studies in Palestine for the US Department of Agriculture, suggested using the waters of the Jordan, Yarmouk, Banias, Hasbani, Dan, and Zarqa Rivers. Lowdermilk advocated combining these resources in a plan to irrigate the Jordan Valley, portions of the northern

Galilee and northern Palestine. Lowdermilk and other American consultants also proposed to divert the Litani River in southern Lebanon to form an artificial lake in the Galilee whose waters would be pumped southward to irrigate the Negev desert. To carry out such super projects, he suggested a Jordan Valley Authority modeled on the Tennessee Valley Authority in the United States. In late 1944 the World Zionist Organization invited James Hays to Palestine to work out the details of Lowdermilk's concepts.

Fed mainly by the diverted Jordan water, Israel's National Water Carrier, as it came to be called, was a large pipe capable of channeling 320 Mm^3 each year from Lake Tiberias to Rosh Haayin, near Tel Aviv. Later, extensions were built to carry water as far south as the Negev desert. Along its course, the Israelis were able to recultivate the areas from which the Palestinians had fled by using more intensive farming methods such as drip irrigation. Cultivated land increased from about 1,600 km^2 (400,000 acres) in 1948 to well over 4,000 km^2 (1.1 million acres) before 1980, nearly half of which was irrigated land.

Beginning in 1953, Syrian artillery shelled the construction and engineering sites behind the town of Tiberias, forcing the Israelis to move the main pumping station. Recognizing the explosive potential of the water dispute, US President Dwight Eisenhower appointed Ambassador Eric Johnston to the formidable task of finding water-sharing arrangements acceptable to all parties. Johnston's proposals were based largely on the work of Charles Main, a Boston consultant. Their basic principle was that water within one catchment area should not be diverted outside that area, regardless of political boundaries, until the needs of those within the catchment area were satisfied. Because diversion of the Jordan River in Israel was already virtually irreversible, the Eisenhower administration resolved to meet the needs, if possible, of both Arabs and Israelis in the catchment area defined by Galilee, southern Lebanon and western Syria. The Johnston Plan involved the diversion of Jordan River water to Israel and was not accepted by the Arabs. The head of the Palestine Truce Supervisory Organization (a U.N. committee set up to monitor the cease-fire) enjoined Israel to cease, but after Israel refused, the United States suspended economic aid to Israel forcing its compliance; aid was resumed in October 1953.

The Johnston Plan included building a number of small dams in Lebanon, Syria, and Israel, as well as a high dam on the Yarmouk River between Jordan and Syria and the draining of the Huleh swamps in Israel. The Johnston Plan allotted Syria 45 Mm^3 of water a year, Jordan 774 Mm^3, and Israel 394 Mm^3. However, Israel and the Arab states objected, asking for larger shares. Nevertheless, Israel quickly completed the draining of the Huleh marshes during 1951–58, and many of the small dams were built. The Yarmouk dam was never constructed. Even though funds have been pledged by a US and World Bank-led consortium for many years, lack of agreement on how to divide the waters collected in the new lake to be formed on the Yarmouk among Jordan, Syria, and Israel was still holding up the start of the main construction work in 1992.

The Arabs also insisted on a Jordan Valley Authority (JVA) and an international supervisory body, while Israel rejected giving a board containing Arab members any control over Israeli water supplies. Like so many Arab–Israeli issues, this one became a major public relations contest in the United States. *The New York Times* columnist Cyrus Sulzberger recorded in November 1953 that "Israel is now using shrewd diplomacy in an effort to have blame attached to the Arab states for turning down . . . the new Jordan Valley Authority . . . the Israelis are very clever and will always have an advantage over the Arabs states on economic know-how. But the Arabs are getting wise to the diplomatic maneuver and therefore are planning a smarter diplomatic game on the JVA business."[8]

For the Arabs, the best thing to come of the Johnston proposals was probably Jordan's East Ghor canal, eventually built largely with American funding and well designed to make Jordan's side of the valley a fertile, California-like setting for orchards and truck gardens which have helped feed Jordan, its neighbors and even Arab Gulf states over the years. However, the Johnston Plan also included a second irrigation canal west of Jordan, on land Jordan controlled until Israel captured it in 1967. This second canal was never built.

Table 3.2   Water allocations to riparians of the Jordan river system

	Water Allocation According to the Unified (Johnston) Plan (Mm^3)			Current Use Levels in 1990 (Mm^3)		
	Jordan River	Yarmouk River	Total	Jordan River	Yarmouk River	Total
Jordan	343	377	720	243	120	363
Syria	42	90	132	0	170	170
Israel	375	25	400	540	100	640
Lebanon	35	0	35	0	0	0
Total	795	492	1287	783	390	1173

As shown in table 3.2,[9] Israel extracts some 100 Mm^3 annually from the Yarmouk although the annual share of the occupied Yarmouk Triangle is only 25 Mm^3. According to the Johnston Plan, Jordan's share from the Yarmouk was to be 377 Mm^3 annually of which 100 Mm^3 were to be stored in Lake Tiberias. However, Jordan only uses 120 Mm^3 annually of this flow due to the other riparians using more than the planned allocation. On the other hand, Israel uses 640 Mm^3 per year of water from the Jordan and Yarmouk, which is 240 Mm^3 higher than envisaged in the Johnston Plan.

Due to shortage in the annual rainfall, Israel has not been able to fulfill its part of the commitment to provide Jordan with 100 Mm^3 annually from Lake Tiberias.[10] The breakdown was never fully overcome until peace was achieved in 1994 between Jordan and Israel. Indeed, the rise of Palestinian military power in Jordan late in the 1960s led to Israeli air attacks in Jordan's East Ghor Main Canal (today's King Abdallah Canal). When the focus of Palestinian military activism shifted to Lebanon after 1970, however, Jordan and Israel gradually managed to achieve a more civilized, if still tense, modus vivendi with respect to water.

From the 1970s onward the principal focus of Israeli–Jordanian water interactions has been the Yarmouk River. As noted below in the discussion of treaty provisions pertaining to the Yarmouk, Jordan, and Israel have respectively been able to use (on average) 130 Mm^3 and 70 Mm^3 annually from this waterway, with the balance of the average annual stream flow either depleted by (something in excess of 200 Mm^3) or lost to the Dead Sea in the form of raging winter floods (about 70 Mm^3).[11]

Over time (starting in the 1980s) Israeli and Jordanian technicians, meeting during the summer irrigation months under U.N. auspices, would adjust flows more or less cooperatively, dredging the river bed and using sand bags to slow and divide the waters. Even so, Israel was adamantly opposed, without legally-binding water allocations, to the damming of the Yarmouk at Maqarin station and to the construction of a diversion weir at Adasiya. In 1980 the United States, represented by veteran diplomat Philip Habib, sought without success to broker an agreement between the parties that would allow Jordan to build the dam and the diversion weir.[12]

Bilateral and Multilateral Negotiations[13]

The Gulf War in 1991 and the collapse of the Soviet Union caused a re-alignment of political alliances in the Middle East which finally made possible public face-to-face peace talks between Arabs and Israelis, in Madrid on October 30, 1991. During the bilateral negotiations between Israel and each of its neighbors, it was agreed that a second track be established for multilateral negotiations on five regional subjects, including water resources. These two mutually reinforcing tracks – the bilateral and multilateral – have led, at this writing, to a treaty of peace between Israel and Jordan and a declaration of principles for agreement between Israel and the Palestinian Authority. Both have had a water component in terms of allocations and projects. In neither has water had any influence on the discussions over final boundaries.

Israel–Jordan Treaty of Peace

On October 26, 1994, Israel and Jordan signed a treaty of peace, ending a state of war that had lasted for nearly half a century. For the first time since the states came in to being, the treaty legally defines mutually recognized water allocations. Acknowledging that "water issues along their entire boundary must be dealt within their totality," the treaty spells out allocations for both the Yarmouk and Jordan Rivers and Arava/Araba groundwater and calls for joint efforts to prevent water pollution. Also, "[recognizing] that their water resources are not sufficient to meet their needs," the treaty calls for ways of alleviating the water shortage through cooperative projects, both regional and international.

The peace treaty also makes some minor boundary modifications. As noted, the Israel–Jordan boundary was delineated by Great Britain in 1922 and followed the center of the Yarmouk and Jordan Rivers, the Dead Sea and Wadi Araba. In the late 1960s and 1970s, Israel had occasionally made minor modifications in the boundary south of the Dead Sea to make specific sections more secure from infiltrators. They had also done so on occasion to reach sites from which small wells might better be developed. In the last sixteen years, no modifications were made except on the rare occasion that one of these local wells ran dry and had to be re-dug. All of these territorial modifications were reversed and all affected land was returned to Jordan as a consequence of the peace treaty, although Israel retains rights to the water which comes from these wells.

Provisions Concerning Both Rivers

Article III (Water Quality and Protection) applies to both the Yarmouk and the Jordan Rivers. The provisions may be summarized as follows:

1. The parties will protect, within their own jurisdictions, both rivers from "pollution, contamination, harm or unauthorized withdrawals from each other's allocations."
2. A Joint Water Committee will establish stations to monitor water quality along the Israel–Jordan boundary.
3. The parties "will each prohibit the disposal of municipal and industrial waste water in the course of the [rivers] before they are treated to standards allowing their unrestricted agricultural use." This prohibition became effective on October 26, 1997.
4. Water supplied by one party to the other from a given spot shall be equivalent in quality to water used by the supplier at that same spot.
5. "Saline springs currently diverted to the Jordan River [by Israel] are

earmarked for desalination within four years. Both countries shall cooperate to ensure that the resulting brine will not be disposed of in the Jordan River or any of its tributaries."

6. The parties will protect water they supply to each other "against any pollution, contamination, harm or unauthorized withdrawal of each other's allocations."

Article V: Notification and Agreement

1. Artificial changes in or of the course of the Jordan and Yarmouk Rivers can only be made by mutual agreement.
2. Each country undertakes to notify the other, six months ahead of time, of any intended projects, which are likely to change the flow of either of the above rivers along their common boundary, or the quality of such flow. The subject will be discussed in the Joint Water Committee with the aim of preventing harm and mitigating adverse impacts such projects may cause.

Article VI: Co-operation

1. Jordan and Israel undertake to exchange relevant data on water resources through the Joint Water Committee.
2. Jordan and Israel shall co-operate in developing plans for purposes of increasing water supplies and improving water use efficiency, within the context of bilateral, regional or international cooperation.

Article VII: Joint Water Committee

1. For the purpose of the implementation of this Annex, the parties will establish a Joint Water Committee comprised of three members from each country.
2. The Joint Water Committee will, with the approval of the respective governments, specify its work procedures, the frequency of its meetings, and the details of its scope of work. The Committee may invite experts and/or advisors as may be required.
3. The Committee may form, as it deems necessary, a number of specialized sub-committees and assign them technical tasks. In this context, it is agreed that these sub-committees will include a northern sub-committee and a southern sub-committee, for the management on the ground of the mutual water resources in these sectors.

Israel–Palestinian Declaration of Principles and Interim Agreement

On September 15, 1993, the "Declaration of Principles on Interim Self-Government Arrangements" was signed between Palestinians and Israelis, which called for Palestinian autonomy in, and the removal of Israeli military forces from, Gaza and Jericho. Among other issues, this bilateral agreement called for the creation of a Palestinian Water Administration Authority.

Moreover, the first item in Annex M of the Interim Agreement, on cooperation in economic and development programs, included a focus on cooperation in the field of water, including a Water Development Program prepared by experts from both sides, which would also specify the mode of cooperation in the management of water resources in the West Bank and Gaza Strip and would include proposals for studies and plans on water rights of each party, as well as on the equitable utilization of joint water resources for implementation in and beyond the interim period.

Negotiations between Israel, Syria and Lebanon

Since 1994, water has been raised in official negotiations between Israel and Syria. Serious bilateral talks have only taken place since the fall of 1995. The basis for Israel–Syria negotiations is the premise of "land for peace," in this case, an exchange of the Golan Heights for peace. The discussions thus far have focused on interpretations of how much Golan and with what security arrangements, for how much peace. The crux of the territorial dispute is the question of to which boundaries Israel would withdraw – the boundaries between Israel and Syria have included the international boundary between the British and French mandates (1923), the Armistice line (1949), and the cease-fire lines from 1967 and 1974.

The Syrian position has been an insistence of a return to the borders of June 5, 1967, while Israel refers to the boundaries of 1923. Although it has not been mentioned explicitly, the difference between these two positions is precisely over access to water resources. The only distinction between the two lines is the inclusion or exclusion of the three small areas which made up the demilitarized zone between 1949 and 1967 – Givat Banias, the hill overlooking Banias Springs; the Daughters of Jacob Bridge area; and the town of Al-Hamma / Hamat Gader – a total of about 60 km^2. Each of these three territories were included in British Palestine specifically because of their access to the Jordan and Yarmouk Rivers and since each is a relatively low-lying area with no strategic importance, their access to water is still considered paramount.

Conclusion [14]

Israel and Jordan have set the stage for an era of genuine cooperation in water matters. Indeed, cooperative professional relationships established over many years between Jordanian and Israeli water experts should greatly facilitate the rapid implementation of those provisions of Annex II of the peace treaty. The parties also have before them some unresolved issues worthy of consideration:

1. International assistance will be essential for the implementation of key provisions. International financial institutions and bilateral donors, notwithstanding the euphoria of peace, may link some, if not all, of their help to the accomplishment of internal water-related reform in both countries. Both parties have long been guilty of considerable waste (e.g., relatively unproductive agricultural uses of water). As Natasha Beschorner has noted, the real water crisis in the Middle East "relates fundamentally to the nature of water allocation and use within states rather than to water allocation between states."

2. Although an Israel–Jordan peace treaty clearly cannot address the influence and interests of a third party, Syria's importance is obvious. A 1987 Syrian–Jordanian treaty gave Jordan Syria's permission to build the Unity Dam across the Yarmouk River, the boundary between the two states. In return Syria was authorized to dam many of the wadis (intermittent streams) feeding the Yarmouk during the winter and to irrigate land within the Yarmouk watershed. Syria's depletion of the Yarmouk's base flow (from springs) and winter runoff exceeds 200 Mm^3 annually – well over 40% of the river's "historical" average annual flow. As matters now stand Jordan is caught in the middle of ever-increasing Syrian depletion and specific, quantitative obligations to Israel. Jordan, which needs the Yarmouk River much more than do its neighbors, has no legally-recognized allocation – neither a numerical quantity nor a percentage – of the river's flow. The expansion of peace in the region to include Syria should, one would hope, inspire an international assistance effort aimed at sharply reducing Syria's depletion of the Yarmouk's headwaters. Pressure pipe and drip irrigation methods, commonly employed throughout Israel and Jordan, coupled with new, less water-intensive cropping patterns, can enable Syria to forgo significant amounts of water in favor of downstream users while still meeting the municipal and agricultural needs of southern Syria.

3. Downstream of both parties lies an emerging Palestinian entity. It is clear that Israel and Jordan took care, during the course of their negotiations, to limit their discussion of the Jordan River to places north of where the West Bank fronts on the Jordan River. Although the rehabilitation of

the Jordan River would be a gift of unsurpassed value to downstream users, it is possible that the expectations of both Jordan and Israel will need to be weighed alongside downstream expectations. Viewed in the context of a relatively short but often violent history, Israel and Jordan deserve to take pride in what they have accomplished. Yet even though they settled their differences without the aid or involvement of outsiders, the implementation of their undertakings now depends very much on external actors, both regional and international. Annex II of the peace treaty is a good beginning, but a beginning nonetheless.

Notes

1   Water and Instability in the Middle East, Adelphi Paper No. 273 (London: Brassey's (UK) Ltd., 1992).

2   For example, in response to Israel channeling water from Lake Tiberias to the central and southern reaches of the country through the canals and pipelines of the National Water Carrier, the Arab Summit Conference meeting in Cairo in 1964 approved plans for the diversion of the headwaters of the River Jordan. Syria and Lebanon began construction of a canal to redirect water from the Banias and Hasbani rivers to the Yarmouk River. Israel responded with long-range tank, artillery, and air attacks that brought the diversion project to a halt and further heightened tensions which erupted two-and-a-half years later in the 1967 war.

3   "Entitlement" as used here refers to a state's legal claim to a specific share of the common water resource. Actual consumption of the resource need not necessarily be limited to a state's entitlement. Through market or other mechanisms – e.g., barter, purchase, or "loans" of water from other user – a state could consume water in excess of its entitlement. The prior definition of entitlements, however, is essential before such distributive mechanism can operate.

4   For detailed description of the Jordan River basin, see N. Kliot, The application of Helsinki and ILC Rules to the Jordan–Yarmouk Drainage Basin, in *Water Scarcity and Conflict in the Middle East* (London: Routledge, 1993), Chapter 3; J. Kolars, Water Resources of the Middle East, *Canadian Journal of Development Studies*, Special Issue (1992): 109–10; and T. Naff and R. C. Matson (eds.), *Water in the Middle East: Conflict or Co-operation?* (Boulder, CO: Westview Press, 1984), pp. 17–22.

5   Source: T. Naff and R. Matson (eds.), *Water in The Middle East: Conflict or Cooperation?* (Boulder, CO: Westview Press, 1984).

6   James H. Moore, Parting the waters: Calculating Entitlements, *Middle East Policy* 3 (2) (1994): 91–95.

7   Much of this historical material has been repeated and updated from J. K. Cooley, The War over Water, *Foreign Policy* (Spring 1984): pp. 3–26.

8   C. L. Sulzberger, *A Long Row of Candles, Memories and Diaries* 1934–1954 (New York: The Macmillan Company, 1969), p. 925.

9   Source: Figures of the Unified (Johnston) Plan are quoted from T. Naff and R. C. Matson, eds., *Water in the Middle East: Conflict or Cooperation* (Boulder, CO: Westview Press, 1984), p. 42; Current use levels in 1990 are

quoted from J. Deason and E. Salameh, The Jordan River Basin: paper presented to the World Bank International workshop on Comprehensive Water Resources Management Policy (Washington, USA, June 24–28, 1991), p. 38.

10   J. K. Cooley, Middle East Water: Power For Peace, *Middle East Policy* 1 (2) (1992): 5–6.

11   F. Hof. The Yarmouk and Jordan Rivers in the Israel–Jordan Peace Treaty. *Middle East Policy* 3 (4) (1995): 47–56.

12   Water Projects in the Peace Treaty: Terms of Reference for Engineering Consulting Services ("Draft No. 1"), The Hashemite Kingdom of Jordan (Brussels, November 28, 1994) p. 2.

13   Much of this material has been repeated and updated from "Water, War, and Arab–Israeli Peace Negotiations," Paper presented by A. T. Wolf, University of Alabama (Bloomington, Indiana, March 7, 1996).

14   Much of this material has been repeated and updated from, N. Beschorner, Water and Instability in the Middle East, Adelphi Paper 273, Winter 1993/1994, p. 3.

CHAPTER 4

Compliance with and Violations of the Unified / Johnston Plan for the Jordan Valley

MUNTHER J. HADDADIN

In June 1980, Ambassador Philip Habib, the United States Special Envoy to the Middle East, expressed in a meeting with Jordanian officials at the home of Jordan's Minister of Foreign Affairs that the Johnston Plan was too old and no longer applicable.[1] On October 25, 1986, the Israeli Water Commissioner, Meir Ben Meir, stated in a seminar organized by the Center for Strategic and International Studies in Washington DC, that the Johnston Plan, in Israel's view, was "archaic, and no longer applicable."[2]

In the sixth round of bilateral negotiations between Jordan and Israel that took place in the United States State Department building under the Middle East Peace Process, I, as the Jordanian negotiator, proposed in late August 1992 that the two sides could adopt an energy and time saving option by falling back on the consensus they had arrived at through the mediation of the United States in 1955; failing that, the negotiations would have to start all over again. Several material and moral gains had accrued to Israel between 1955 and the 1992 negotiations: it had dealt Egypt a military defeat in the Sinai Campaign of November 1956; had stunned the Arabs by a blitz war in June 1967 and occupied Sinai, the Gaza Strip, the West Bank of Jordan and the Golan Heights of Syria; had received the President of the largest Arab state, Egypt, in Israel in 1978; had concluded a peace treaty with Egypt in 1979; and had occupied Southern Lebanon in 1978 and invaded that country as far north as its capital, Beirut. Naturally, it would not be attractive for Israel to return to a 1955 formula and the Israeli response to the Jordanian proposal was cold; its lawyer-negotiator showered the Jordanians with questions about that consensus, the Johnston Plan.[3]

Was the Plan Rejected?

The Johnston Plan, known officially as the Unified Plan for the Development of the Jordan Valley, was forged by Ambassador Eric Johnston, President Eisenhower's Special Envoy to the Middle East. Johnston engaged in shuttle diplomacy between Israel and the Arab riparian parties: Lebanon, Syria, and the Hashemite Kingdom of Jordan, which included the West Bank at the time. Israel formed a committee headed by the Foreign Minister, Moshe Sharett, who later became Prime Minister during the Johnston shuttle; the Arab League, upon Egypt's initiative, formed an Arab Technical Committee in December 1953 to negotiate the technical aspects of the Jordan Valley Plan with the Johnston Mission.[4] Johnston conducted four rounds of separate talks with Israel and the Arab riparian parties in an attempt to reach a consensus on a development plan to harness benefits from the utilization of the Jordan River waters by all the riparian parties. Through these negotiation rounds, both the Arab Technical Committee and Israel effected negotiated amendments to the first draft of the Plan that Johnston started out with. The final version of the Plan was finalized in September–October 1955.[5] Essentially, the Plan specified water shares for the riparian parties, demarcated storage reservoirs, and called for international supervision of the implementation. Johnston insisted that, once the riparian share was defined, the riparian state was free to use the share the way it saw fit. That note by Johnston was meant to allay Arab objections to the transfer of water outside the Jordan basin, and gave a green light for Israel to use its share, or parts thereof, outside the Jordan basin to irrigate the Negev and make more room for settlement of Jewish immigrants expected to come from Russia and Europe.

The utilization of the Jordan waters was to take place before the river exit from Lake Tiberias, and the utilization of the waters of the tributaries south of the Lake, including the Yarmouk River, was to take place before the waters discharged into the Lower Jordan. In effect, this would transform the Lower Jordan (south of Lake Tiberias) into an agricultural drainage watercourse with practically no fresh water inflow except what floods might spill over from Lake Tiberias and the lower tributaries in good rainy years. This aspect of the Plan was not environmentally sound, especially that Israel took advantage of it later and managed to dispose untreated wastewater and divert saline springs water into the Lower Jordan.[6]

On the water shares issue, only the irrigation needs of arable lands in the river basin were taken into account. Municipal and industrial needs of urban and rural areas in the basin were not accounted for. These needs proved to be substantial as time passed by. Additionally, lands classified as non-arable because of topography were reclassified into arable lands when pressure pipe distribution systems were adopted in the 1970s.

Contrary to opinions expressed by distinguished writers[7] in the water literature, the Johnston Plan was NOT rejected by the Arab states and Israel. On the Arab side, Johnston, in a meeting in Beirut with the Foreign Ministers of Jordan, Syria, Lebanon, and the Egyptian Ambassador to Syria, reached agreement with them on the elements of the Plan and documented it in a memorandum dated 19 February 1955. The meeting was chaired by the Lebanese Prime Minister and was attended by the Syrian Prime Minister in his capacity as Foreign Minister.[8] That agreement was amended at the expense of the water share of the Hashemite Kingdom of Jordan in order that Johnston would secure Israel's approval of the Plan. Despite their dismay at that amendment, the Arab Technical Committee approved the Plan as amended and recommended, in late September 1955, that the Political Committee of the Arab League approves it. Johnston lobbied for the Plan with Prime Minister Jamal Abdul Nasser of Egypt, who suggested that the Plan be divided into two parts, economic and political.[9] He was sure the Arab League ministers would approve the technical part, and that, in due time, they would consider the political part.

In their meeting, 8–11 October 1955, the Arab League Political Committee, composed of the foreign ministers of the eight member countries, decided to endorse a decision taken by the four concerned states: Lebanon, Syria, Jordan, and Egypt. Their decision was "despite the efforts exerted to produce the Plan, certain important points still need further consideration. It was therefore decided that the experts be asked to pursue the mission with which they have been entrusted until an agreement safeguarding Arab interests is reached."[10] The experts the decision referred to were the Arab experts of the Arab Technical Committee and their supporting technical experts. The decision of the Arab League political committee was communicated to Ambassador Johnston on 15 October through a letter signed by the League's Secretary General, Mr. Abdul Khaliq Hassouneh. The position of the Arab League was therefore not that of rejection of the Plan but was that of requesting more studies of it in order that Arab rights be secured.

The Israeli position on the Plan was described in a cable to the US Secretary of State from Tel Aviv dated October 14, 1955. The cable indicated that Ambassador Johnston's talks with the Prime Minister, Foreign Minister and their staffs went very well. The two ministers had accepted the technical side of the Plan. In view of the uncertain status of the Plan on the Arab side, the Israelis did not plan formally to submit it to their new cabinet at that time. The cable further stated that, "it thus appears that informally both Israel and the Arab States have reached a common basis on a water settlement formula and that the sole but formidable remaining obstacle is to obtain Arab concurrence at the political level."[11]

It is apparent that neither the Arab side, nor the Israeli side "rejected" the Plan. It is equally clear that they both accepted its technical part, but were hesitant to accept the political part.

The political part that disturbed the Arab side, as expressed by the Syrian Prime and Foreign Minister, Sa'eed el Ghazzi, was the potential cooperation, albeit indirect, with Israel, a state they never recognized and with which they were in a state of war. The Plan, they feared, would entail an implicit recognition of Israel before the underlying grievances of the Arabs were addressed. Such grievances were rooted in the right of Israel to exist as a foreign state on Arab lands, the fate of the Palestinian refugees and their right of return, the Palestinian property taken over by the state of Israel, and other issues. Domestically, the Syrian Parliament debated the issue before the Cairo meeting of the Political Committee of the Arab League and recommended that any Syrian government should not accept the Plan. Similarly, the Lebanese Parliament decreed a similar recommendation, but the Lebanese delegation came to the Cairo meetings "with open hearts and minds, and that they would take back with them whatever decisions the ministers made and try to convince parliament," as expressed by the Lebanese Minister, Salim Lahoud, at the Cairo meetings.[12] However, the position of the government and not that of either parliament was the position that mattered.[13]

Developments in the region were not conducive to the resumption of Johnston's talks for final approval of the Plan. In early 1956, the hawkish David Ben-Gurion replaced the dovish Moshe Sharett as Prime Minister of Israel; Premier Jamal Abdul Nasser nationalized the Suez Canal in July of the same year, a move that upset both Britain and France as major shareholders in the Suez Canal Company; and the build up for the Suez invasion by Israel, Britain, and France began. The political atmosphere was hardly conducive to talks towards any unified plan for the development of the Jordan Valley in which Israel would be a party.[14]

Later in the 1960s, Israeli Prime Minister Levi Eshkol aired Israel's official position on the Johnston Plan at a press conference on 10 July 1963 in which he assured Israel's adherence to the provisions of the Johnston Plan short of the international supervision on withdrawals. Such a declaration was made upon insistence by the United States.[15] Again, on 22 January 1964, Prime Minister Eshkol said, "Israel would pump from Lake Tiberias within the limits of quantities allotted to it under the Johnston Plan."[16]

Conformity with the Plan

The United States was intent on having the riparian parties commit to the Unified Plan. Soon after Johnston reported his findings to his superiors the Plan was forwarded in January 1956 via diplomatic channels to all the riparian parties and to Egypt. The primary elements of the Plan are contained in Appendix I.

The United States had further appeared as a fair party in the wake of the Sinai Campaign during which it ordered the forces that occupied the Sinai (Israel), and parts of the Suez Canal (Britain and France), "in violation of the United Nations Charter," to withdraw their forces. The Arabs developed trust and a liking of the United States which began to move into the region and replace the colonial powers (Britain and France) that had been there previously. Jordan, one of the riparian parties, terminated its treaty with Britain in December 1956 and soon became a recipient of American military and economic assistance. Israel, on the other side, was hoping for continued American financial assistance to help build the economy of a new state and make room for more Jewish immigrants. The political developments in the region[17] resulted in the United States coming to Jordan's assistance, and it agreed to finance the development of the Jordanian part of the Jordan Valley under the East Ghor Canal Project. The Project was in the East Jordan Valley and drew water from the Yarmouk River, the major tributary of the Jordan. The United States, however, saw to it, before the assistance was provided, that Jordan would not draw from the Yarmouk River more flows than allocated to it under the Unified/Johnston Plan.[18] To Israel's dismay, the United States refused to have a private understanding with Israel, as a concerned downstream riparian, over the Jordanian project in advance for fear of word spreading that the Jordanians had a secret deal with Israel, a matter that would have embarrassed the Jordanians immensely at the time.

On the Israeli side, several exchanges of notes with the United States over an Israeli project using Jordan River water took place between 28 January 1958 and 7 January 1960. The exchanges and meetings culminated in US support of Israeli projects in the amount of $15 million on condition that documented assurance be provided by Israel that its project was consistent with the Johnston Plan.[19]

The United States thus imposed the conditions on the riparian parties to have the Johnston Plan's provisions adhered to as they intended to use waters from the Jordan River or its tributaries. To ensure adherence and the fulfillment of each to the given commitments, the United States kept a watchful eye on the projects of both riparian parties, Jordan and Israel. Frequent visits by Dr. Wayne D. Criddle, the American expert who was the technical arm of Eric Johnston, were made to the region during which he inspected the works on both sides and submitted reports to the State Department.

The above clearly shows that the United States remained faithful to the implementation of the Plan it had engineered and sponsored. It used its influence to have the parties conform to the provisions of the Plan, at least as far as the water shares were concerned. The other riparian parties, Lebanon and Syria, did not start any projects to utilize water from the Jordan basin, and no American finance was therefore required. Consequently, no commitments by either of the riparian parties were ever made to adhere to the Plan.

Violations of the Plan

The Unified Plan, among other provisions, called for the use of Lake Tiberias/ Sea of Galilee as a common storage reservoir in which the Jordan and Yarmouk waters, in excess of the current demands, would be stored for use in drier months by Israel and Jordan. The other storage provision was a dam on the Yarmouk at Maqarin where the average annual flow was about half the average flow of the Yarmouk at Adassiyya. The United States hoped that a common water reservoir in Lake Tiberias would eventually trigger cooperation between the Arabs and Israel. Israel, however, did not favor that arrangement nor did the Arabs like it, each party for their own reasons. On its part, Israel feared that the storage of Arab waters in Lake Tiberias could prompt them to claim sovereignty over it, and the Arabs were fearful of storing part of their waters in a lake falling totally under the jurisdiction of their enemy.[20] As no agreement on the issue of storage of Yarmouk floods in Lake Tiberias was reached, the Plan called for deferment of this part of it for five years and made it contingent upon a decision of an international board of expatriate engineers who would decide to adopt it or otherwise store the excess Yarmouk floods in another location that could be more economical.

Even before Ambassador Johnston started his mission, Israel had begun the construction of its National Water Carrier that would transfer Jordan River water to the Negev. The construction started from the Negev in the south towards the north. The intake of the Carrier, originally designed to be immediately downstream of Jisr Banat Ya'coub on the Jordan north of Lake Tiberias, was shifted to the northwest corner of the lake for technical and security reasons. The Carrier was scheduled for completion in 1963. When Israel announced it would commence pumping from Lake Tiberias into the National Water Carrier in late 1963, Nasser of Egypt called for an Arab Summit to discuss the matter.[21] The Arab leaders met in Cairo in January 1964 and decided to have the upper tributaries of the Jordan, flowing within their lands, diverted to Lebanon, Syria, and to the

Hashemite Kingdom of Jordan. The subject tributaries were the Hasbani in Lebanon, and the Banyas in Syria. Such a measure would rule out the provision of storage in Lake Tiberias that the United States favored. Additionally, Israel feared that such diversion could draw more water than the Plan allocated to both Syria and Lebanon. The diversion project, however, would have left waters from both tributaries, especially winter flows, to flow to Israel and Lake Tiberias with no claim by the Arabs to any of it. Additionally, the Arab diversion would have left a sizeable spring flow originating downstream of the Yarmouk diversion, estimated at 80 million cubic meters (Mm^3) per year, to continue downstream towards Israel. This would increase Israel's share from the Yarmouk stipulated in the Plan at 25 Mm^3 per year. All in all, the Arab diversion project would not have secured the total Arab shares as stipulated under the Johnston Plan. The author estimated the total diversions to be around 600 Mm^3 compared to a total Arab share of 644 Mm^3 under the Johnston Plan.[22] The Arab diversion of the Banyas and the Wazzani stalled in 1966, and the construction of the Mukheiba dam in Jordan, a key component of the Arab diversion project, had to be interrupted after the northern bank of the Yarmouk was overrun by the Israelis as they occupied the Golan Heights in the June war of 1967.

The violations of the Plan commenced shortly after the June war of 1967. On Israel's part, its diversion of Jordan waters from Lake Tiberias increased gradually after 1964 and reached Israel's planned diversion of 320 Mm^3 in the early 1970s. However, the Arab shares of Syria, Lebanon, and the West Bank in the upper Jordan River, all totaling 267 Mm^3 per year, are still in Israeli hands as are most of the lands that were to depend on these waters for irrigation.[23] Israel has been using these shares in its own territories and these waters are to form part of the topics of negotiations between Israel and the other Arab parties when the Peace Process eventually is resumed.

On the Yarmouk the violations were more serious; Israel was using its share and more since the mid-1960s. It installed pumps on the river at Baqura and felt free to pump as much water as it could, especially during the winter floods. It used these waters concurrently and stored the excess in Lake Tiberias for later use. By 1990, the total Israeli use from the Yarmouk reached about 90 Mm^3 per year as compared to its share under the Plan of 25 Mm^3. This Israeli violation was checked and rolled back during the Middle East Peace negotiations, and was documented in Annex II to the Jordan–Israel Peace Treaty of 26 October 1994.[24]

Syria, for its part, initiated an irrigation development program in the eastern Golan–western Horan regions of the Yarmouk catchment. Despite a treaty concluded with Jordan in 1953, Syria started to build dams on the Yarmouk tributaries in its territories to impound floodwater earmarked for storage in the Maqarin dam for Jordan's benefit. The first such dam was the Der'a dam built in 1967, and the number increased to become 42 dams by the year 2000 with a total storage capacity around 170 Mm³. The Johnston Plan allocated to Syria from the Yarmouk a net depletion of 52 Mm³ per year corresponding to 90 Mm³ withdrawals and 38 Mm³ return flow, and that allocation of withdrawal, instated upon Syria's insistence in 1954, was consistent with the Jordan–Syria treaty over the Yarmouk. Additionally, Syria increased its use of the spring waters in the Yarmouk catchment and substantially increased the drilling of wells through which abstraction of the ground water feeding the Yarmouk was highly increased. The estimated Syrian use of the Yarmouk is about 220–230 Mm³ per year compared to its share of 90 Mm³. It is noteworthy that the summer flow of the Yarmouk at Adassiyya dropped from about 6100 liters per second in August of 1963 to less than 2000 liters per second in the summer of 2000.

In the Hashemite Kingdom of Jordan, irrigation development of the Jordan Valley started with the US supported East Ghor Canal Project in 1959 and was stalled in 1967 by the turbulence associated with the June war and its aftermath. The Yarmouk was more than sufficient to fully irrigate some 11,400 hectares in the East Jordan Valley by 1967. A project to irrigate about 3500 hectares in the West Bank from the Yarmouk via the East Ghor Canal was aborted as a result of the war of 1967. Jordan resumed its irrigation development in the East Jordan Valley in 1973 (the West Bank fell to Israeli occupation in 1967). The primary water source was the unregulated flow of the Yarmouk River augmented with stored water in reservoirs it had built on the side wadis of Wadi al Arab and the Zarqa River, all within Jordan's rightful share under the Johnston Plan. Jordan was also counting on Syria's adherence to the 1953 treaty over the Yarmouk, and drew its development plans accordingly. However, Jordan became aware of Syria's ambitious irrigation program in Yarmouk catchment and had been aware of Israel's objection to Jordan's intent to build a diversion structure across the Yarmouk to enhance its ability to divert its share from the Yarmouk. Pressured by Syrian violations upstream and Israeli violations downstream, all at Jordan's expense, Jordan's development plans for the Jordan Valley faced hardships and are still facing them because of upstream abstractions of groundwater and surface water in the Yarmouk catchment. However, Jordan was able to build a diversion dam across the Yarmouk and to check Israeli violations as a result of its bilateral negotiations with Israel.[25] Estimates of Syrian withdrawals from the Yarmouk run at about 220–230 Mm³, highly in excess of its Yarmouk share under the Johnston Plan of 90 Mm³. The excess in Syrian uses is at the expense of the share of the Hashemite Kingdom and the share of the West Bank. Under the Johnston Plan, the share of the Hashemite Kingdom (East Bank and West Bank then) would amount to 377 Mm³ from the Yarmouk assuming only 52 Mm³ depletion by Syrian use. Following the bases on which Johnston allocated the shares (arable areas and irrigation duty) the author, as Jordan's water negotiator in the Middle East Peace Process, computed the share for each side of the Kingdom. Such a split in shares became all the more necessary as a result of the legal and administrative disengagement from the West Bank, a decision taken by Jordan in 1988. Jordanian negotiators were instructed not to indulge in negotiations on behalf of the Palestinians; they had to mind their own issues of conflict with Israel. The 377 Mm³ share was thereby split into 296 Mm³ for the East Jordan Valley (Hashemite Kingdom) and 81 Mm³ for the West Jordan Valley (West Bank). The share of the Hashemite Kingdom in Lake Tiberias, set at 100 Mm³ with no more than 15 Mm³ of brackish water, was all part of the share of the West Bank to be delivered to it via the East Ghor Canal through a siphon under the Jordan River at Deir Alla.[26] The West Bank share from the Lake Tiberias was to be delivered to Jordan's East Ghor Canal by gravity flow in a carrier canal connecting the lake with the East Ghor Canal at a designated drop near Adassiyya.

Jordan was unable to secure its full share in the Yarmouk, mainly because of the increased upstream use by Syria, and partly because it could not build the Maqarin dam and the associated diversion weir at Adassiyya. Israel's insistence to have a say in the project and to secure a higher share than that allocated to it in the Johnston Plan was one reason behind stalling the project; the other reason was Syria's insistence that Jordan agree to conclude a new treaty with it over the Yarmouk by which Jordan conceded to Syria the right to the 26 dams it had built on the Yarmouk tributaries up to 1987. Jordan abided by the Syrian demands in 1988; it also negotiated the diversion dam with Israel in 1994 and had it built in place by 1999.

As for the proposed storage in Lake Tiberias, Jordan was able to find an economically viable site where Yarmouk floods can be stored and proceeded to build the Karama dam, an off-river storage facility in the Jordan Valley. It was almost filled with Yarmouk waters in the 1997–1998 season.[27] Its capacity is 55 Mm³ at a cost of $82 million. During the implementation of the peace treaty between Israel and Jordan, Israel agreed to reserve a total of 60 Mm³ of storage capacity in Lake Tiberias for use by Jordan, as will be elaborated below.[28]

As for the upper tributaries, the Dan flows totally within Israel although the aquifer that feeds it lies primarily in Syrian and Lebanese lands, and the Banyas has been under Israeli occupation since 1967. There has been little use by Lebanon of the Hasbani for irrigation until recently (late 2002) when Lebanon decided to withdraw water from the Wazzani springs for municipal purposes. Lebanon was well within the share allocated to it under the Johnston Plan but Israel raised serious objections that increased tensions in the region. The United States sent its water experts and quieted things down as Lebanon proceeded with its project. This conflict is detailed below.

As for Lake Tiberias, which today is the only surface water reservoir in Israel, the transformation of it into a reservoir through the Degania gates at the exit of the Jordan River from it created complications that became one of the stumbling blocks in the Syrian–Israeli track of negotiations (elaborated below.)

To summarize: the visible violations of the Johnston Plan have been committed by Syria and by Israel. The first exercised its violations in the Yarmouk catchment both on surface water and groundwater, and the second committed its violations by using the shares of Lebanon, Syria, and Palestine (the West Bank) in the upper tributaries. Syria's violations run at about 140 Mm^3 at the expense of Jordan and Palestine, and Israel's violations run at about 167 Mm^3 at the expense of Lebanon (25 Mm^3), Syria (42 Mm^3), and Palestine (100 Mm^3).

By 1980, when Ambassador Philip Habib expressed the opinion that the Johnston Plan was no longer workable, he probably based his judgment on the above violations. And when the Israeli Water Commissioner, in October 1986, described the Johnston Plan as "archaic," it was not without the interests of Israel placed up front. Israel definitely would lose if the Plan were to be fully implemented.

Does the Plan Need Adjustment?

The Unified/Johnston Plan was worked out by the Johnston Mission through strenuous negotiations between antagonistic riparian parties who could not conduct direct negotiations. The Arab parties at the time did not recognize Israel, nor was Israel close to recognizing the rights of the Palestinian people. Key developments since then have changed those attitudes. The first such development was the outcome of the 1967 war and Security Council Resolution 242 of 22 November 1967 which gave implicit recognition of Israel by the Arab states that accepted the resolution (Syria and Iraq did not). The second development was the emergence of the Palestine Liberation Organization (PLO) and its assumption of the responsibility as the sole legitimate representative of the Palestinian people.[29] The third key development was the 1973 war between Israel on one side and Egypt and Syria (assisted by troops from Jordan and Iraq) on the other. Syria indirectly accepted Resolution 242 when it accepted Security Council Resolution 338 of 1973. A turning point was in 1979 when Egypt concluded a peace treaty with Israel. Additionally, Israel overran Lebanese territory as far north as Beirut in 1982, and Iraq invaded Kuwait in August 1990 but was evicted by a coalition led by the United States in January–February 1991. Subsequent to this latter event, the United States launched a Middle East Peace Process in Madrid on October 31, 1991 in joint sponsorship with the Soviet Union.

The Arab core parties and Israel engaged in peace talks in 1991 under the Middle East Peace Process in which direct bilateral negotiations were conducted. Separately, the Palestine Liberation Organization conducted secret bilateral talks with Israel that yielded the Oslo Accords in September 1993. The legal restrictions against bilateral contacts between Israel and the PLO were lifted. Jordan concluded a peace treaty with Israel in 1994 that contained an annex detailing a resolution of the water conflict. As things stand now, there is no prohibition against direct talks between Israel and any of the riparian parties with whom the Jordan waters are shared.

On the resource–population front, a drastic imbalance occurred in the water resources–population equation. The per capita share of renewable water resources has dropped drastically since 1955 due to immigration, population shifts, and improvement of the standard of living. The Jordan basin, despite the high costs incurred in capital and in operations, became the source of municipal water supply to urban areas in the region (Amman, Jordan's capital, Irbid in North Jordan, and towns in Lebanon), a fact that was not taken into consideration when the Johnston Plan was being negotiated. The per capita income of the riparian countries, and of the region at large, made strides especially after the oil boom in 1973. The pumping for irrigation, viewed unfeasible in the mid-1950s, was adopted in the late seventies. The generation of power, viewed valuable in water schemes, gave way to steam power plants and to gas turbines. The technology of drilling for groundwater assumed an advanced scale never before practiced, and wells have been drilled in the Jordan watershed that have impacted the base flow of the tributaries of the Jordan. Environmental awareness and concerns have intensified manyfold over what they had been at the time the Plan was formulated.[30] Agricultural technology and irrigation technology made leaps that enabled an increase in agricultural yields per unit flow of water; agricultural research and extension today play a more prominent role than they did in the 1950s. And, finally, there will be no DMZs (demilitarized zones) after a comprehensive peace is achieved in the region, and no contested lands on either side of the borders.[31]

The negotiators of the Plan did not give importance to soil water in the basin or in the territories of the riparian countries; nor did it pay attention to the disparity in national incomes and development needs. True, the irrigation duties in the northern part of the basin were higher in the Plan formulation than the drier southern parts, but that was due to the amounts of evapo-transpiration in both regions rather than an adjustment for soil water in the not-so-dry north. Additionally, the Plan did not address ground water abstraction that could affect the base flow of the tributaries or the river itself, or its availability elsewhere in the countries of the riparian parties. Finally, although the Plan specified annual shares for each riparian party, it did not specify the locations in the basin where abstraction was to be allowed with the exception of the Adassiyya diversion dam on the Yarmouk and the diversion structure north of Lake Tiberias (for Israeli diversions); it did not specify the rates (hourly or daily) of such abstraction.

There also were no provisions in the Plan for drought management. Furthermore, some provisions of the Plan have been superseded; an example is Israel's diversion structure on the Jordan whereby this diversion has been changed to the northwest corner of the Lake Tiberias at Tabgha. Other provisions need not be adhered to, such as the use of the East Ghor Canal in Jordan to convey Lake Tiberias water to the West Bank via a siphon under the Jordan. Other minor provisions can be ignored, such as the power generation systems on the East Ghor Canal.

The mitigation of the adverse environmental impact of the Plan should be negotiated with focus on the Jordan River itself below Lake Tiberias and on the Dead Sea. Jordan and Israel, under the auspices of the Trilateral Economic Committee,[32] agreed to build jointly a project to control the level of the Dead Sea at elevation –396 meters below sea level. This should largely mitigate the adverse environmental impact due to freshwater diversions in and out of the Jordan River basin, and due to evaporation from the industrial pans the two countries have built for the extraction of minerals from the Dead Sea.[33] A third concerned riparian, the future Palestinian State, is yet to join the scheme. In addition to the visible impacts, the invisible impact is more serious. The lowering of the Dead Sea is causing the ground water aquifers in the mountains to the east and west to drain more groundwater into the sea.

Another important aspect of the Plan is the adoption of Lake Tiberias as a common reservoir. Israel proceeded in 1963 to transform the Lake Tiberias into a surface water reservoir by building a gated structure at the exit of the Jordan River from the lake. Further, it has invested in the improvement of the quality of water in the lake by diverting the saline springs downstream.[34] By virtue of the gated structure at Degania, the upper water level of the Lake was raised about 4 meters and thus its surface expanded horizontally and, at different reservoir levels, infringed upon occupied Syrian territory.[35] This, the author thinks, triggered another dispute, a territorial one, in the Syrian–Israeli talks that collapsed in March 2001. Syria insisted that all its occupied territories be returned, including the territories adjacent to the northeast shore of the Lake; Israel, on its part, was adamant in insisting that the entire surface of the Lake (including the inundated Syrian lands) be totally within Israeli sovereignty.

The above developments and changes, the most notable being the drastic transformation of the nature of relations between the Arab states and Israel, warrant a serious comprehensive look at the basin and call for a sober, in-depth treatment that will result in a positive sum arrangement with gains for each of the riparian parties. The environmentally adverse impacts, in particular, should be mitigated, especially those affecting the Jordan River course and the Dead Sea. The two water bodies should be considered important parts of the world heritage and international efforts should be launched to rehabilitate their deteriorating status. Several other aspects of the Johnston Plan can be amended as briefly outlined in Appendix I.

Compliance with the Plan in the Peace Talks

The outcome of the water negotiations between Israel and Jordan conformed to the provisions of the Johnston Plan although no reference was made to it. Israel's share of the Yarmouk was specified at 25 Mm^3 per year just as the Plan stipulated. Details were added (a) to locate the point of diversion to Jordan, (b) to build a diversion weir across the river at Adassiya, and, (c) to specify the amounts due to Israel in the dry months (15 May–15 October), and during the rest of the year.

An important adjustment was made to the Plan by including the needs for municipal water in the basin; this was set at 50 Mm^3 of "additional water of drinkable quality" due to Jordan. Another provision was included by which a mutual concession was made by the two countries: Israel was allowed to pump 20 Mm^3 of Yarmouk winter flow (from Jordan's share) in return for 20 Mm^3 of water that Israel would supply from its share in the upper Jordan to Jordan during the dry months from Lake Tiberias.[36] This has been working well since 1995 to the advantage of both parties.

A subsequent agreement was signed between the concerned ministers of both countries on March 10, 1998.[37] The agreement provided the detail for building the diversion weir, and for the storage of 60 Mm^3 of Yarmouk floods in Lake Tiberias for Jordan's benefit.[38] While the construction of the weir was accomplished in December 1999, the diversion of floodwater to Lake Tiberias in the order of 60 Mm^3 has not materialized, primarily because of the political tensions prevailing in the area emanating from the collapse of negotiations on the Israeli–Palestinian track and the eruption of the Palestinian Intifada. However, it is believed that Israel is pumping the winter flow of 20 Mm^3 from the Yarmouk, allowed under the treaty, primarily for storage in Lake Tiberias, and the agreement of March 10, 1998 is clear in its stipulation to have Jordan benefit from Lake Tiberias as a storage facility. Under these arrangements, the use of Lake Tiberias as a common reservoir, and the use of the Karama Dam as an economic storage location for the Yarmouk floods, both envisaged by the Unified/Johnston Plan, have actually materialized in recognition of its validity.

The peace talks between Israel and the other riparian parties collapsed in 2000 with the Palestinians, and with Lebanon and Syria. Israel, however, effected a unilateral withdrawal from Lebanon in May 2001 with the exception, Lebanon says, of the Shaba'a farms.

In 2002, Lebanon initiated a domestic water project by which it would withdraw water from the Wazzani springs that feed the Wazzani tributary to the Hasbani to serve the population of villages in the area. The waters of the Hasbani, including the Wazzani, flow naturally into the Jordan and Lake Tiberias. Israel was not duly alerted to the Lebanese intentions because the two countries are still in a state of war. Israel reacted with threats of military retaliation to stop Lebanon from going ahead with that project.

The reason behind Israel's anger was, in the author's estimation, twofold: one was the failure of Lebanon to notify Israel of its intentions, and the other is related to the location where Lebanon planned to abstract the water. The author does not believe that Israel denies Lebanon the right to its share in the Hasbani; rather, it was important for Israel to take early note of Lebanon's intention and adjust its own water budget accordingly. Moreover, Israel may have demanded that the Lebanese share be drawn from the Hasbani itself and not from the Wazzani as mentioned above. The reason is as follows: The Lebanese share in the Hasbani, set at 35 Mm^3 per year, was made at the insistence of the Arab Technical Committee during its negotiations with Ambassador Johnston in his second round in June 1954.[39] All the allocations under the Johnston Plan were made to satisfy the *irrigation* requirements of the basin arable lands (by gravity) in the territories of the respective parties. There are no arable lands to speak of that can be irrigated by gravity from the Wazzani tributary. The author's guess is that Israel believes that Lebanon's share should be abstracted from the Hasbani where the potential arable lands lie, and not from the Wazzani. The difference between the two tributaries is in the water quality, which is superior in the Wazzani. So, to conform to the Unified Plan allocations, Israel may believe that Lebanon should not touch the Wazzani, but could use its allocated share from the Hasbani. Lebanon, however, went ahead despite the Israeli threats to implement its domestic water project, with about 12 Mm^3 annual capacity, from the Wazzani springs. Thanks to the intervention of the United States the situation was kept under control.

Conclusion

Neither the Arab side nor the Israeli side rejected the Unified Plan that Ambassador Eric Johnston had worked out through shuttle diplomacy (October 1953 – October 1955). The role of the third party, the United States of America, was instrumental not only in formulating the Plan but also in enforcing its provisions to the maximum extent possible.

Violators of the Plan's provisions have been Israel and Syria, the former on the Yarmouk and on the Jordan and the latter on the Yarmouk. The bilateral negotiations between Jordan and Israel brought about Israel's conformity with the plan provisions on the Yarmouk, but the Plan's violations by Israel on the Jordan and by Syria on the Yarmouk await meaningful negotiations between the parties. Syrian violations on the Yarmouk tax the shares of Jordan and the West bank, and Israel's violations on the Jordan are at the expense of Lebanon, Syria, and the West Bank. One can imagine the extended negotiations that are needed to settle all accounts of allocations.

Finally, the Plan warrants negotiated adjustments to account for the developments and changes that have occurred since the Plan was formulated. One does not expect the violations of the Plan to be accommodated, but the advent of technological advances, demographic shifts, municipal water needs, and, importantly, the environmentally adverse impacts, have to be factored in the water management of the Jordan basin.

Successful outcomes materialized from the bilateral negotiations between Jordan and Israel, and adjustment of the Plan was possible to the advantage of both parties. It is hoped that a similar result through future negotiations will guarantee gains for all the riparian parties.

Appendix I:
Primary Elements of the Unified/Johnston Plan

Storage

A.   The Maqarin Reservoir

The Plan envisions the construction of a dam 126 meters high on the Yarmouk near Maqarin to impound 300 million cubic meters (Mm^3) of water for irrigation and make possible the generation of some 150 million kilowatt hours of electric energy. The Arab parties may increase the storage volume to their liking (460 Mm^3) but the United States will not contribute to the cost of the additional height.

B.   Lake Tiberias

The Plan contemplates the storage of Yarmouk flows in Lake Tiberias. Averaged out over a period of years, these flood flows will amount to approximately 80 Mm^3 a year. Allowing for variable annual stream flows and needs for irrigation, it has been determined that storage space for about 300 Mm^3 will be required in Lake Tiberias. The Plan would assure Arab states of space up to this amount when needed.

The Plan thus proposes to provide the total storage space required for Jordan's irrigation needs through the construction of 300–Mm^3 reservoir on the Yarmouk and through the utilization of approximately 300 Mm^3 of storage capacity in Lake Tiberias.

C.   Deferred Use of Tiberias

The use of Lake Tiberias is deferred for five years. At the end of this

period, the neutral Engineering Board would determine the necessity of
storing Yarmouk flood waters in Lake Tiberias or whether more feasible
and economical storage might be found elsewhere.

D.   The Hasbani
An immediate survey to obtain hydrologic and land use data in the
water shed of the Hasbani River in Lebanon would be undertaken with
funds provided by the United States in the amount of $250,000. The infor-
mation thus obtained would be used to determine the necessity of
constructing a storage dam on the Hasbani to assure that water allocated
for Lebanese lands could actually be made available.

Distribution

Once stored, the waters of the Valley must be conveyed, under careful
regulation, to lands they are to irrigate. The Plan therefore contemplates
the installation of the following facilities to transport water to Arab areas:

a)   A diversion dam near Adassiyya to supply the East Ghor
     Canal and, if necessary, to divert the excess flood waters
     to Lake Tiberias for later delivery to Jordan;
b)   A main canal network in Jordan, including:
     1. The East Ghor Canal running from Adassiyya southward to
        the vicinity of the Dead Sea.
     2. A siphon or other structure for conveying water from the East
        Ghor Canal to the West Ghor.
     3. The West Ghor Canal in Jordan feeding from the siphon.
     4. A feeder canal from Lake Tiberias to a junction with the East
        Ghor Canal.
     5. A canal from Adassiyya to Lake Tiberias, if necessary, to
        capture and store Yarmouk flood flows in the lake.
c)   A distribution system to convey water from the main Ghor canals
     to the farm lands;
d)   Pumping plants to raise water to lands above the main Ghor
     canals;
e)   Generating plants on the main canals to supply power for
     pumping;
f)   Main drainage facilities for removing excess water and salts from
     irrigated lands;
g)   Regulating and control works on Lake Tiberias if the lake is used
     to store Yarmouk flood flows;
h)   A new diversion structure and canal from the Jordan River to
     Boteiha Farm in Syria, together with 50 kW of electric power to
     replace water power;
i)   A diversion structure north of Lake Tiberias to permit Israeli
     withdrawals from the upper Jordan;

j) A main canal from this upstream diversion to irrigable areas in the Galilee hills;

k) A short canal from Lake Tiberias down to the west side of the river to serve irrigable lands in the Beisan area in Israel.

Division of Water

To Lebanon:	35 Mm³ from the Hasbani River
To Syria:	20 Mm³ from the Banyas
	22 Mm³ from the Jordan to irrigate the Boteiha Farm
	90 Mm³ from the Yarmouk
To the Hashemite Kingdom of Jordan:	720 Mm³ as follows:
	243 Mm³ from side wadis and wells as follows:
	175 Mm³ from eastern side wadis
	8 Mm³ ground water, East Bank
	52 Mm³ from side western wadis including 14 Mm³ from Fara'a
	8 Mm³ ground water
	377 Mm³ as the residual flow in the Yarmouk
	100 Mm³ from the Upper Jordan, with up to 15 Mm³ from less-saline springs
To Israel:	The residual flow from the upper Jordan after deducting the shares of Lebanon, Syria, and Jordan.
	25 Mm³ from the Yarmouk
Evaporation (deductible from the shares):	300 Mm³ from Lake Tiberias
	14 Mm³ from the Maqarin reservoir

Supervision

The Plan proposes the creation of an impartial Engineering Board together with a Water Master for the purpose of supervising operation of the water system and compliance of the parties. The Engineering Board would consist of three eminent engineers who would be selected from a list prepared by the Secretary General of the United Nations. One would be selected by the participating Arab states and one would be selected by Israel. The two engineers so selected would choose a third who would serve as chairman.

Notes

1 The author was among the officials present at the meeting. The late Ambassador Habib was mediating between Jordan and Israel to enable Jordan to proceed with the construction of a diversion dam associated with the Maqarin dam to be built on the Yarmouk River with United States financing.

2 This two-day seminar was chaired by Dr. Joyce Starr. The author was a participant delegated by his government.

3 Munther J. Haddadin, *Diplomacy on the Jordan: International Conflict and Negotiated Resolution* (Boston: Kluwer Academic Publishers, 2002), pp. 299–301. See also, Negotiated Resolution of the Jordan–Israel Water Conflict, *International Negotiation* 5 (2000): 263–88.

4 Jordan and Syria had concluded a treaty between them to harness the Yarmouk River in June of that year to facilitate the construction of a dam on the Yarmouk. The water sharing under the Johnston Plan was made in conformity with the provisions of that treaty.

5 Israel contested its share in Johnston's allocation of the Yarmouk water, set at 25 Mm^3 per year, and demanded that her share be set at 40 Mm^3. It cited a memorandum her delegates submitted to Johnston in New York on July 5, 1955, a memorandum the Americans considered reflected only Israel's viewpoint.

6 The subject was brought up in the bilateral negotiations between Jordan and Israel in the group on water, energy, and the environment. Later, the author made a visit to the site in September 1994 when the negotiations were held at Beit Gabriel on the shore of Lake Tiberias.

7 This opinion was expressed by Stephen C. McCaffrey; see his Water Disputes Defined: Characteristics and Trends for Resolving Them, in *Resolution of International Water Disputes* (The Permanent Court of Arbitration/ Peace Palace Papers, vol. 5) (Boston: Kluwer Law Publishers International, 2003). McCaffrey cites Miriam Lowi, *Water and Power: The Politics of a Scarce Resource in the Jordan River Basin* (Cambridge: Cambridge University Press, 1993), ch. 4.

8 See Haddadin, *Diplomacy*, pp. 97–100.

9 Johnston had dinner at Nasser's invitation in the home of the latter at Mainsheet el Barky, Cairo. See a memorandum of conversation entitled "Meeting Between Ambassador Johnston and Prime Minister Nasser, Cairo, October 8, 1955," State Department Records, the United States National Archives, Washington DC. Text reproduced in Haddadin *Diplomacy,* pp. 116–17.

10 See the letter addressed to Ambassador Johnston from the Secretary General of the Arab League, Mr. Abdul Khaliq Hassouneh, dated October 15, 1955, cited by Haddadin in *Diplomacy*, p. 121.

11 Cable No. 356, October 14, 1954, 7 p.m. from US Ambassador in Tel Aviv, Mr. Lawson, to the Secretary of State with copies to Johnston who was in Rome on his way home, to U. S. embassies in London, Paris, Cairo, Damascus, Amman, and Beirut. Copy of the cable contained in Haddadin, *Diplomacy*, pp. 123–25.

12 See Haddadin, *Diplomacy*, p. 120.

13 Note, for example, the contemporary position of the United States Congress and that of the Administration concerning moving the U. S. Embassy from Tel Aviv to Jerusalem which entails recognition of Jerusalem as Israel's capital. Congress voted more than once for the transfer of the Embassy and the Administration refrained from abiding.

14 Minister Mahmoud Riyadh who participated in the Johnston's talks with the Arabs, and succeeded Assonet as Secretary General of the Arab League, told the author in 1984 that Johnston was determined to return to the region with a wider mandate: to try and find an acceptable solution to the Arab–Israeli conflict. "The Suez Campaign shattered his plans," Riyadh said.

15 Haddadin, *Diplomacy,* p. 156.

16 BBC, "Eshkol's Statement on the Cairo Conference," no. 1459, January 22, 1964, pp. A/1–2; cited by Lowi, *Water and Power.*

17 In April 1957, King Hussein aborted a military coup that would have aligned Jordan with Egypt and Syria and the Soviets. In February 1958 Egypt and Syria formed the United Arab Republic and the new union was antagonistic to Jordan. In July 1958, a military coup killed the King of Iraq, a cousin of King Hussein of Jordan, and the rest of the royal family of Iraq. A leftist republic was established there that allied itself with the leftist Arabs, primarily Egypt and Syria, for a short while. An Arab Cold War was on.

18 Note in memorandum no. 58/14/6719 dated February 25, 1958 from the Foreign Minister of Jordan acknowledging and approving an Aide-Mémoire sent to him by the Chargé d'Affaires of the United States dated 24 February 1958. For text see Haddadin, *Diplomacy*, pp. 143–44.

19 Haddadin, *Diplomacy*, pp. 146–47.

20 The Arab Technical Committee, upon the recommendation of Dr. Sushi Mazlum of the Syrian side, accepted the storing of the floods that enter the Yarmouk downstream of Maqarin in Lake Tiberias. The estimated flow was in the order of 90 Mm^3 per year. Israel continued to object to the use of Lake Tiberias for storage of Arab waters.

21 Only one Arab Summit had been convened theretofore, viz. in 1946 at Incas, Egypt, to discuss the dangers befalling Palestine before the creation of the State of Israel.

22 Composed of 120 Mm^3 from the Hasbani–Wazzani, 100 Mm^3 from Banyas and 380 from the Yarmouk, and these are optimistic figures. The Johnston Plan allocated to Lebanon 35 Mm^3 from the Hasbani; 132 Mm^3 to Syria from the Banyas, Yarmouk, and the Jordan; and 100 Mm^3 from Lake Tiberias to Jordan including a ceiling of 15 Mm^3 brackish water, and 377 Mm^3 from the Yarmouk. The total was 644 Mm^3 per year excluding some 16 Mm^3 for evaporation from the Maqarin reservoir.

23 These lands are located in the Hasbani Valley (including the Shaba'a farms) that was under Israeli occupation between 1978 and 2001; the Banyas gorge in the Golan Heights of Syria and the Buteiha farm around Lake Tiberias; and the West Jordan Valley between Wadi Yabis/ Tirat Zvi and the Dead Sea. The Shaba'a farms, the Golan, and the Palestinian Jordan Valley are still under Israeli occupation.

24 See Haddadin, *Diplomacy*, pp. 517–18.

25 A full account of the negotiated settlement is found in Haddadin *Diplomacy.*

26 For details see Michael Baker, Jr. Inc and Harza Engineering Company, *Yarmouk–Jordan Valley Project, the Hashemite Kingdom of Jordan: Master Plan Report,* 8 vols. (Rochester, PA, 1955).

27 King Hussein inaugurated the dam completion in April 1998 before his departure to the Mayo Clinic in July. It was the last project the King inaugurated before he passed away in January 1999.

28 For details see Haddadin, *Diplomacy*, pp. 430–37.

29 That role was made official by the decision of the Arab Summit in Rabat in 1974.

30 This is particularly relevant to the Plan, which made the Jordan below Lake Tiberias an agricultural drain through the water off-takes by the various riparian parties. Israel even uses it for discharging wastewater effluent. It also is significant for the level of the Dead Sea, which has been lowered some 24 meters since the Plan was formulated, with the dangerous phenomenon of *sink holes* on both shores of the Sea.

31 Ambassador Eric Johnston was instructed by the Secretary of State to attempt to adjust the Armistice lines between Syria and Israel to give Syria a part of Lake Tiberias and to eliminate the DMZs.

32 The Trilateral Economic Committee was set up in October 1993 with the United States as Chair, and Israel and Jordan as members. Its task was to identify joint projects that the two states could pursue after peace is reached.

33 That Plan was submitted in a written proposal the author made and presented to the Jordan Government in February 1994, and was submitted to the Trilateral Economic Committee in February 1994. It proposes the integrated development of the Jordan Rift Valley with a canal linking the Dead Sea with the Red Sea at Aqaba as a backbone. The Plan was approved and the two countries agreed to proceed with its studies and preparation of Tender Documents.

34 Most of the saline spring water was diverted into the Jordan River south of the Lake. In the Jordan–Israel treaty, these saline spring waters are to be desalinated and the sweet water allocated to Jordan.

35 See Munther J Haddadin, Water in the Middle East Peace Process, *The Geographical Journal* 168, no. 4 (December 2002) pp. 324–40.

36 See Annex II "Water Related Matters" to the Jordan–Israel Peace Treaty available from the website of King Hussein of Jordan: <http://www. kinghussein.gov.jo/documents.html>.

37 The ministers were Ariel Sharon, Minister of National Infrastructure accompanied by the signature of Meir Ben Meir, Israel's Water Commissioner, and Dr. Munther J. Haddadin, Minister of Water and Irrigation of Jordan.

38 For details see Haddadin's, *Diplomacy,* pp. 434–39.

39 In his first version of the Plan that he distributed to the riparian parties in October 1953, Johnston did not allocate any share to Lebanon.

CHAPTER 5

Water Resources Scarcity in West Africa: The Imperatives of Regional Cooperation

AONDOVER TARHULE

Outside of the Middle East, there are almost no examples in modern times in which wars have been fought explicitly over water resources. Even so, access to and control of water resources has featured prominently in many conflicts around the world. The problem is that the role that competition over scarce water resources plays in many conflicts is often difficult to discern. This chapter identifies salient dynamics in water scarcity and conflicts in semi-arid West Africa. The premise is that improved recognition and appreciation of the ways in which water scarcity may induce or propagate conflicts could help mitigate or even prevent future conflicts. Semi-arid West Africa provides excellent context for exploring this thesis because the economies and livelihoods of a majority of the people are directly and intricately dependent on environmental resources, of which water is the most important. The potential for conflict is analyzed at both local (sub national) and international scales. The policy implications of the dynamics identified are discussed and African initiatives relevant for other water-stressed regions of the world are highlighted.

West Africa is defined geographically and politically as comprising the 16 member-countries of the Economic Community of West African States (ECOWAS; figure 5.1). The population of West Africa as a whole was 247[1] million in 2000, with Nigeria contributing about 53% of that total. Lacking major physiographic features, the regions' geography is closely associated with its climate. Thus, it is customary to divide West Africa into a southern humid (Subequitorial–Guinean) ecological zone and a northern semi-arid

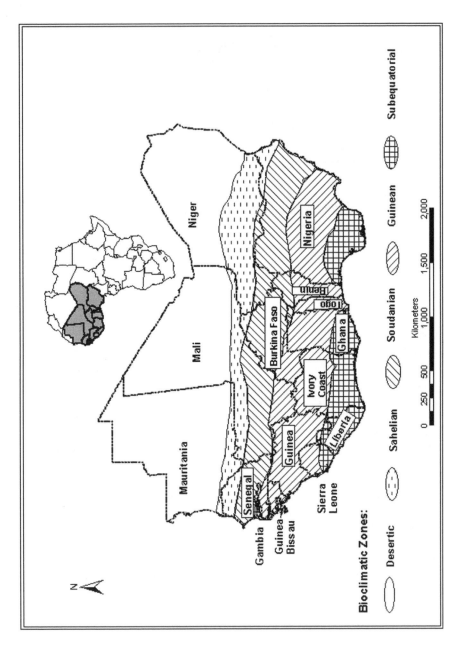

Figure 5.1 Member states of the Economic Community of West Africa (ECOWAS) and the Bio-Climatic Zones Superimposed.

(Soudano–Sahelian) to arid (desert) ecological zone. These zones reflect south–north gradients in rainfall: tropical rain forests along the coasts give way to grass-dominated savanna landscapes that progressively diminish to the scantiness of the Great Sahara desert.

The semi-arid Soudano-Sahelian zone, where problems associated with water scarcity are most severe, is the focus of this essay. This zone has been inhabited for millennia and once supported large African empires including Ghana, Mali and Songhai (the last being the largest indigenous empire formed on the African continent) as well as numerous lesser kingdoms and city-states. Today, the Soudano–Sahelian region is virtually synonymous with drought, reflecting the significance of long-term climatic change.[2] More recently sharply decreased rainfall conditions beginning around 1968 ushered in drought conditions that persist to the present day. At the peak of the drought in 1973/74, between 100,000 to 250,000 people are estimated to have starved to death, along with 12 million cattle.[3]

Six countries – Senegal, Mauritania, Mali, Niger, Burkina Faso, and Chad – lie entirely or substantially within the Western Soudan Sahel, which, in point of fact, extends the breath of the African continent. These countries had a combined population of 57 million in 2000[1] and average growth rate of 2.91%. The vast majority of these people (approximately 75%) lives in the rural areas and engages predominantly in rain-fed agriculture, pastoralism and agropastoralism. Significantly, water – *the* limiting factor in all these activities – is largely unregulated in rural West Africa. As a result, the livelihoods of a majority of the people are intricately dependent on the availability of water supply from natural sources including rainfall, which is highly erratic in space and time. Over centuries of trial and error, life has generally adapted to these conditions. However, new challenges have made some of these coping strategies untenable, with serious implications for conflict. For example, cattle herders traditionally migrated towards the relatively humid south and greener pastures during the dry season, easing the pressure on land and water resources.[4] Recently, however, agricultural expansion and other development activities have choked these migratory routes, compelling herders to spend longer or entire dry seasons in the Soudan Sahel. Similarly, population growth and natural climatic variability during the past three decades have combined to put intense pressure on West Africa's water resources.

This chapter reviews the effects of these developments on the water resources in the region, the potential for conflict as well as the range of possible technical solutions. The focus is on three major issues – River Niger, Lake Chad, and drought – which epitomize the nature of water scarcity and the associated problems in West Africa. Each of these is described below, including the nature of the problem, areas of successful cooperation as well as the outstanding sources of concern.

River Niger

With a drainage basin that covers fully 10 countries, the 4,200 km long River Niger is the life-blood of the western Soudano–Sahelian region. The original Tuareg name for this river, "egerou n-igereou" meaning "river of rivers," bears eloquent testimony to its historical and cultural significance. As the only perennial river in the West African Sahel, River Niger delimits a corridor of agricultural productivity that offers refuge to the people of the area even in times of desolate drought (figure 5.2). Indeed, for most of its flow through the countries of Mali, Niger, and, to a lesser extent, Nigeria, the River Niger offers the only feasible means for major irrigation, domestic and industrial water supply and hydropower generation. But diminished rainfall throughout the Soudano–Sahelian region (figure 5.3a) has led to dramatic reductions in the flow of the River Niger (figure 5.3b). The discharge, measured at Niamey (Republic of Niger) decreased from about 1050 $m^3 s^{-1}$ during the 1950s and '60s to about 500 $m^3 s^{-1}$ in the mid-1980s before recovering to around 750 $m^3 s^{-1}$ in the later 1990s. Figure 5.3 represents these decreases as normalized departures from the long-term (1940–1998) average stream discharge. At Jebba (Nigeria), the average annual flow of the river after 1980 is about 40% less than the flows prior to 1980. Such drastic changes in the major water resources of the area have the potential to disrupt the socio-economic and agro-ecological systems that depend on it. It is important therefore to assess the risk, and identify potential sources, of conflict among the countries sharing the drainage basin over the use of the River Niger's water resources. The type, magnitude and patterns of resource use, as well as the agreements that guide the use of that resource among the stakeholders, is critical in inducing conflict or fostering cooperation. Below, I review the water use patterns on the River Niger to provide insight into the conflict potential.

Despite its significance to subsistence livelihoods in the Sahel, the River Niger is grossly underutilized when measured against most indices. For example, in almost all regions, agriculture invariably is the largest user of water. But agricultural water use in the Niger Basin is very limited. The estimated irrigation potential within Mali and Niger is about 556,000 ha and 140,000 ha respectively.[5] However, only about 130,000 ha (Mali: 24%) and 16,000 ha (Niger: 11%) are actually irrigated. These countries are important and emblematic because they comprise 25% each of the total drainage area of the River Niger basin. Such low level of agricultural water use has resulted in minimal demands on the river's water resources serendipitously minimizing the potential for conflict despite severe reductions in flow magnitude.

Hydroelectric power development and the attendant damming of rivers is also a major cause for conflict between upstream and downstream

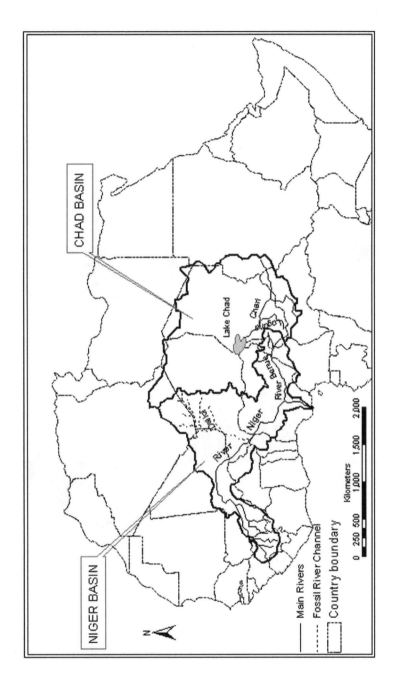

Figure 5.2 The Drainage Basins of the River Niger and Lake Chad.

countries in shared river basins. There are two relatively large hydroelectric power projects on the River Niger, both of which are located in Nigeria, a downstream country. These are the 760 MW capacity Kainji station and the 560 MW Jebba plants.[6] In contrast, there is no major hydroelectric project on the river in the upstream countries. Two irrigation diversion dams exist in Mali at Sotuba (downstream of Bamako) and Markala (near Segou) but these are generally small.

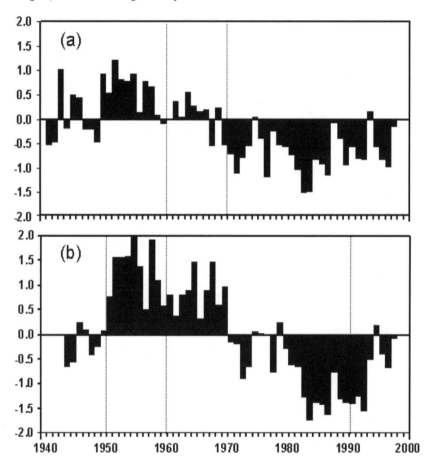

Figure 5.3   Time Series (1941–2000) of Average Normalized **(a)** April–October rainfall departure (σ) for 20 stations in the West African Soudano–Sahelian zone (11°–18°N) east of 10°W. Re-normalized and updated from earlier versions in Lamb (1985) and Lamb and Peppler (1992), where further details can be found; and **(b)** annual stream flow discharge for the River Niger at Niamey (Station Number 1321500048, 13° 40′ 56″ N, 02° 05′ 10′ E).

Looking down the road, larger dams on the River Niger are inevitable because economic development (however the term is defined) in the Sahelian countries hinges critically on increased utilization of water for irrigation and hydroelectric power generation. To fully appreciate this imperative, it is important to keep in mind that Niger and Mali are the second and tenth least developed countries in the world, respectively, according to the United Nations human development Index (<www.undp.org/hdi>). Presently, only 8% of the populations of Mali and Niger have access to electric power. In the rural areas, the proportion of inhabitants with access to electric power is a mere one percent. As a result, urban and rural populations alike depend heavily on wood as a fuel for cooking, which is already contributing to problems of deforestation and desertification in some areas.[7] These problems threaten the livelihoods of the people of the area, and demands for changes in regional energy policy have become more urgent.

Available evidence suggests that both countries (Mali, Niger) recognize the need for boosting their power generating capacity as a prerequisite for economic development and as a means of combating deforestation. Mali, for example, has completed feasibility studies for building a major dam at Tossaye, while Niger has estimated the hydroelectric power potential at Kandaji (near Tillaberi) at 235 MW (or 1330 GWh).[8] Based on such potential, Niger resolved since the late 1970s to build a dam for hydropower generation at Kandaji. However, such upstream dams could potentially compromise the viability of downstream projects, especially during periods of persistent multi-year droughts as previously described. Major irrigation schemes may also further degrade the water quality exacerbating the already serious problem of water hyacinth.

Nigeria stands to lose the most from these developments. The country has developed ambitious plans to expand its own hydropower generation capacity. Currently, only 40% of Nigeria's urban dwellers and 10% of its rural inhabitants have access to electricity supplies. Although the country has an installed capacity of 5900 MW, actual production is on the order of 1500 MW, due in part to decreased flow on the River Niger. Nigeria plans to expand its electric generation, transmission, and distribution systems, with the long-term goal of reaching 25,000 MW in generating capacity by 2010.[9] Although the Nigerian plans incorporate plans to diversify hydropower development away from the River Niger, the Kainji project is likely to remain pivotal to the national electricity grid system and the country is unlikely to view upstream interference with its operation favorably.

To forestall attempts by Niger to build its own hydropower plants, Nigeria supplies 40% of Niger's electricity demand from the Kainji dam. The problem is that electricity consumers in the Republic of Niger pay substantially higher tariff (about 100% more) per unit of power compared to Nigerians, prompting persistent calls for a review of the tariffs.[9] Nigeria also insists on payment in US dollars, another sore point with Niger, and motivation to continue work on its proposed dam at Kandaji. There is a risk therefore that the situation could generate into conflict at some point in the future, unless the countries involved take active steps to develop mutually beneficial strategies. An article in *The Guardian*,[10] arguably Nigeria's most reputable newspaper, illustrates the problem. Titled "Hostility looms as Niger plans to dam River Niger" the article quotes an official from Nigeria's Ministry of Power and Steel:

> Over the years, the country (Niger) has always told us that they wanted to dam the river but each time we persuaded them not to do so, while we ensure that they enjoy concessionary rates from us. But now, the government of the country has taken a definite step and we will have to take measures to protect Nigeria and its electricity supply . . . the situation will create problems between the countries . . .

In point of fact, the ten countries that share the Niger basin recognize the need for cooperation and integrated development of the watershed. In 1964 the Niger Basin commission was constituted to oversee the integrated development of the basin in a manner that both ensures sustainability and minimizes conflict. In November 1980, the Niger Basin Authority (NBA) replaced the Niger Basin Commission as the overseer of regional treaties and conventions. Several of the treaties governing the management of the watershed, as well as the activities of the NBA, are available online http://www.abn.ne/studies.html. The NBA works closely with other agencies with Sahel-wide mandates including the permanent interstate committee on drought and desertification (better known by its French acronym CILLS), the Africa Center for Meteorological Applications for Development (ACMAD), and the Center for Agricultural, Hydrological and Meteorological Research (AGRHYMET). The NBA will be severely tasked once upstream countries can afford to, and insist upon, unilaterally building their own dams. This is significant given the imbalance in military strength between Nigeria (the major downstream user) and the upstream countries. But the prospects for the upstream dams are dim because of the low domestic power market and the high initial investment capital required for the construction of dams, the installation of engineering equipment, and the transmission of power. In other words, the present peace appears largely due to the inability of Niger and Mali to attract investment capital for their dams, rather than to good neighborliness or the effectiveness of the NBA.

Even so, the existence of a collaborative framework is encouraging and provides a model for addressing problems as they emerge. The NBA must be judged by its present achievements, not its perceived future weaknesses. And on this criterion, other regions can benefit from the experience of the NBA in shared watershed management.

Lake Chad

As mentioned previously, climatic variability rather than anthropogenic overuse is considered the driving factor for diminishing water resources in semi-arid West Africa. Lake Chad exemplifies this process. The lake has shrunk considerably during the past three decades due to climatic fluctuation, increased desertification within its drainage basin, and anthropogenic water use. The surface area of the lake in 1963 was approximately 25,000 km, but by July 2001 it was only 1,350 km,[11] a mere one-twentieth of its former size. The lake is very responsive to changes in rainfall because annual inflow accounts for 20–85% of the lake's volume. Ninety percent of Lake Chad's water flows in from the Chari (Shari) and Logone River systems from Cameroon. About 5 percent is contributed by the Komadougou–Yobe system in Nigeria. Beginning with the onset of drought in 1968, annual discharge from the Chari River decreased by 75 percent and the Komadougou–Yobe River, which was perennial, now flows only during the rainy season. Recent modeling results by Coe and Foley[11] revealed that Lake Chad shrank by 30 percent between 1966 and 1975. Comprehensive information concerning the economic losses and the ecological imbalance caused by the lake desiccation is hard to get. But a sense of the problems can be gleaned from the fact that the lakeshore has retreated by more than 100 km in the northern and western parts. The economic loss to the fishing industry and irrigation agriculture can only be imagined.

Anthropogenic water use, principally irrigation, accounted for only 5% of that decrease, with natural climatic variability accounting for the remainder. However, between 1983 and 1994, irrigation demands increased fourfold, accounting for 50% of the additional decrease in the size of the lake. The South Chad Irrigation Project (SCIP) based in Nigeria is the largest of these irrigations schemes. SCIP planning started in 1962–63, the very peak of the wet years, and was based on the unusually high lake levels of the 1950s. A system of pumps and canals was installed to carry water from the lakeshore intake point to farmers' inland fields. However, between 1974 and 1984, the project operated for only six years due to low lake levels, and not since then. The problem is expected to worsen in the coming years as population and irrigation demands continue to increase because, with a drier climate and less rainfall, agricultural areas become more desperate for water to irrigate their crops and will continue draining what is left of Lake Chad.

Despite these changes, major conflict has not yet occurred over the use of Lake Chad's water resources. Two factors account for this situation. First is the recognition or perception among basin inhabitants that climatic variability, not anthropogenic overuse or mismanagement, is to blame for the desiccation of the lake. Right or wrong, this perception has created a sense of shared tragedy, suppressing sentiments that could otherwise lead to conflict.

The second factor is the existence of Lake Chad Basin Commission (LCBC), created by the Fort Lamy (now N'Djamena) Convention signed on 22 May, 1964, by the Heads of State of the four countries which share the lake, namely Cameroon, Niger, Nigeria, and Chad. In 1994, the Central African Republic was admitted as the fifth member state.[12] The LCBC has provided avenues for political, technical, and economic cooperation at the highest levels. The problem here is that as the lakeshore retreats further into the Republic of Chad, the residents of that country may come to assume a proprietary claim over the waters of the lake irrespective of any standing multinational agreements to the contrary. At the present time, the borders of Niger and Nigeria effectively no longer lie within the lake. The second problem – and one already listed by the LCBC as a major challenge – is that environmental refugees (former fishermen and farmers) deprived of their source of livelihoods as the lake retreats have begun to migrate en masse to the urban areas and to cross the borders into neighboring states. It is inevitable that some of these would turn to a life of crime, straining relations among the countries involved.

Finally, as water becomes scarcer, there is a tendency for water managers to hold back more water in reservoirs to compensate for operational deficits. Therefore, it should be expected that if drought conditions persist, upland users might hold back or divert even more water from the rivers feeding into Lake Chad, effectively accelerating its demise. This could lead residents in the vicinity of the lake to protest the water deprivation, leading to regional conflict.

Drought

In a comprehensive review of the impact of drought on the economies of Sub Saharan African countries, Benson and Clay[13] note:

> the dearth of investigative studies appears to reflect the fact that drought has been typically perceived as a problem principally of agriculture and, in partic-ular, food supply with economies automatically and immediately restored to their long-term growth paths with the return of the rains. (p. 2)

The same criticism could be applied with respect to the linkage between drought, or water scarcity in general, and conflict. Drought or acute and widespread water scarcity could induce conflict through migration, when large numbers of otherwise productive and independent people are compelled to abandon their livelihoods for uncertain futures frequently in societies that are ill equipped to absorb them. Between 1981 and 1985, 150 million Africans (about one-third of the population of the continent at the time) were threatened by drought that forced 20 million to abandon their homes in search of food and water.[14]

Furber[15] and Tarhule[16] have attributed the Tuareg Rebellion of the 1990s to droughts occurring in 1973/1974, and again in 1984/1985, which decimated the livestock of the Tuareg and severely imperiled their traditional pastoral lifestyle. With their livelihoods wiped out, the Tuareg were forced into a life of destitution in shantytowns in Algeria, Libya, and Nigeria and in the capital cities of Niger (Niamey) and Mali (Bamako), where they were treated as second-class citizens. Desperation and humiliation arising from this situation fomented resentment against the central governments in Mali and Niger – the two countries in which majorities of the Tuareg reside – and fueled demand for political autonomy. Ultimately, the seven-year long rebellion consumed nearly three thousand lives and generated more than 120,000 refugees.[15] The Tuareg rebellion highlights the salient point that the scarcity of water is usually a source, but rarely the cause, for conflict. But recognition of the source is essential to mitigating future conflict. Additionally, because of the time lag between the initial perturbing factor and the final outcome, as well as the intervening processes, the role of water scarcity in contributing to conflict is not always obvious.

Yet, despite massive and frequent drought-induced migrations in Africa, there are relatively few outbreaks of major conflicts, such as the Tuareg rebellion, that could be attributed to environmental refugees. There are two reasons for this. The first has to do with a "staggered" or "displacement migration" pattern common during periods of drought or environmental stress. As an environmental disaster unfolds and residents of one area migrate out, other groups from environments facing even more severe stress subsequently claim environments abandoned by the first group. For example, the drought-induced famines of 1983–85 displaced around 300,000 people in Northern Sudan, mainly from the Dafur district. In place of the retreating Sudanese, around 110,000 migrants from the areas between Abeche and Arde in Chad flocked into Darfur district.[17] Meanwhile, over 300,000 Eritrean and Ethiopian drought refugees also arrived in Eastern Sudan. The Ethiopian refugees would eventually number in excess of 500,000 before the crisis was over. Here, the Ethiopian refugees took over areas already abandoned by the Sudanese, which reduced the potential for conflict between the two groups.

A second factor mitigating drought-induced conflicts is that most refugees tend to head to areas that are culturally and linguistically similar, allowing them to blend in with the local populations. This was the case during the 1973/74 drought when an estimated 55–80% of the populations of some villages in Niger migrated into Sokoto State in Nigeria.[18] But Southern Niger and Northern Nigeria are ethnically and linguistically homogeneous, inhabited by the Hausa Fulani, and the perception of kinship could not have hurt in facilitating such mixing. But even this cama-raderie could change if worsening economic conditions produce high unemployment in the host communities or if the risk of communicable diseases, such as AIDS, continues to escalate to the point where countries begin to restrict massive influx of people into their borders. Also, the risk for conflict increases once refugees end up in areas where such similarities and cultural ties do not exist.

The most direct cause of conflict occurs when people converge on a few viable water sources. In West Africa, riverine swamps and floodplains have become major focal points as droughts have exposed the food insecurity of the region. Many upland farmers that practiced mainly rain-fed agriculture have moved to the floodplains to adopt irrigated agriculture utilizing water from the shallow alluvial aquifers. Herders, too, have begun to settle permanently in the vicinity of major floodplains to graze their cattle over the rich grass during the dry season. This development has pitched herders against farming communities in the Soudano–Sahelian zone. Violent encounters from increasingly well-armed sides have become frequent, with the number of fatalities rising. In Western Mali, 12 people were killed in violent confrontation between herders from Mauritania and farmers in northern Mali over water use at a well.[19] In Niger, seven people were killed and 43 injured in clashes between Fulani herders and Zarma farmers in the arrondissements of Tera and Birni N'Gaoure, near the riparian village of Falmaye, 90 km southeast of Niamey.[20] These events conform to the diffused, persistent, subnational conflicts hypothesized by Homer-Dixon and Howard.[21,22] Clearly, it is imperative that West Africa as a whole increases its capacity for dealing with water scarcity.

Conclusion and Recommended Solutions

This chapter has discussed the dynamics and examples of water-induced conflicts in semi-arid West Africa. The conclusion that emerges is that cooperation is imperative in shared river basins. Sooner or later, unilater-alism leads to conflict. This simple fact must permeate the political calculus of leaders in areas where water resources are scarce. Indeed as many people have observed, the scarce water resources in the Middle East provide opportunities for fostering cooperation among the warring factions. The Middle East could benefit from successful examples of cooperative initia-tives. For West Africa, the Arab–Israeli conflict is a stark reminder of the imperatives of regional cooperation. Perhaps more so than other parts of the world, West Africa has a unique opportunity to address water issues in an integrated manner before economic and political developments close the window of opportunity. This could be achieved by doing several things:

(1) Improved Scientific Understanding

Sub-Saharan African countries and especially the vulnerable countries in the Soudano–Sahelian zone must improve their understanding of both the causes and implications of water scarcity. Such understanding must begin with a comprehensive inventory of the region's water resources and an assessment of the current and anticipated future climatic and anthropogenic demands on the various stocks of water. This information could then be used to develop national water strategies. A review of the status of water by the Working Group on Water Supply and Sanitation Collaborative Council of the UN concluded that the single most important factor behind water scarcity in Africa is poor water policies. In addition to articulating new policies, there is also a need to reform existing water policies, including the acknowledgment of water as a scarce resource and its centrality to poverty reduction, economic growth, and conflict.[23]

Better scientific information is needed on the long-term patterns of climatic variability, which affects water availability. For example, there is general agreement that both the magnitude and persistence of the present Sahel droughts are unprecedented since the beginning of scientifically measured data. However, it is still unclear whether such mega droughts are a recurring feature of the Sahelian climate or whether the current drought represents a new phenomenon in the climate dynamics of the region.[24] The answer to such questions is critical for future water resources planning, especially for long-lived projects as well as for the allocation of existing water resources such as Lake Chad and the River Niger.

Finally, there is need also for improved understanding of the impact of water scarcity on various aspects of the Sahel community. For example the impacts of drought on the economies of the region are still poorly understood. The linkages and pathways between water scarcity and conflict must be identified at various temporal and spatial scales. Because water scarcity per se is rarely *the* cause of conflict, it would be especially important to assess how water scarcity interacts with other stress factors to induce conflict. This information, combined with the inventory of water resources, could be used to create an interactive water resources and conflict database that could be accessed by researchers and policymakers to manage the effects of water scarcity.

(2) Private Investment

Economic development hinges fundamentally on the availability of electricity. Given their poor and declining economies, the countries of the Sahel will never be able to provide this essential commodity to their citizens. It is imperative, therefore, that the countries of the region cooperate. For example, the prospects for the proposed dam at Kandaji would improve significantly if Burkina Faso, Mali, and Niger collaborated on the project rather than Mali independently attempting to build its own dam at Tossaye. To help attract foreign investment for such a venture, the countries must also open up the hydropower market to private investment. Despite the low purchasing power, the size of the market as well as antici- pated future growth should make the economic returns on private investment attractive.

Concomitant with efforts to attract foreign investment must be attempts to increase public spending in water resources research and development. This is critical because there is some evidence that some water related prob- lems are actually caused by bureaucratic failures or technological limitations in exploiting and distributing available water resources.

(3) Renewable Energy Resources

The regional cooperative apparatus that currently exists needs to explore alternative sources of power supply such as solar and wind energy. However, both the technical expertise required and the initial capital investment in research and infrastructure exceed what the individual countries could afford. Regional cooperation provides a means for over- coming these obstacles. The countries of the region must carefully weigh the cost effectiveness (economies of scale) of cooperative ventures on the one hand, and the potential risk of conflict as a result of unilateral action on the other. At the present time, even the basic data required to make preliminary assessments of the feasibility of alternative energy sources does not exist. Mali, for instance, does not have a national electricity grid and the development of wind power has been considered as a possible solution for outlying villages. However, no data on wind velocities and duration are available.

(4) Conflict Mediation

In the final analysis, conflict at various scales is unavoidable. It is important therefore to set up mechanisms for conflict mediation and mitigation in preparation for when (not if) conflicts arise. Towards this goal, a database of conflict as a resource for scientific research and development planning is an essential first step. Political scientists and security scholars can browse, query, or study such databases to reveal causes, trends, patterns, and connections that may otherwise not be obvious. Research based upon such empirical data would be more comprehensive and persuasive than those based upon isolated case studies or, worse, back of the envelope esti- mates. Politicians and policymakers can utilize the results of such studies to weigh the cost and benefit of various courses of action versus the conse- quences of inaction.

Finally, the institutional capacity of the existing regional organizations, such as the NBA and LCBC, must be strengthened through adequate funding and legislation to prepare them to deal with current and anticipated problems.

Notes

1 <www.cia.gov/cia/publications/factbook/>, 2002.

2 S. E. Nicholson, Climatic Variations in the Sahel and other African Regions during the Past Five Centuries, *Journal of Arid Environments* (1978): 3–24.

3 M. H. Glantz (ed.), *Drought Follows the Plow* (Cambridge University Press, 1994), p. 197.

4 Lloyd Timberlake, *Africa in Crisis: The Causes, the Cures of Environmental Bankruptcy* (London: International Institute for Environment and Development, 1985).

5 FAO, *Irrigation Potential in Africa: A Basin Approach*. FAO Land and Water Development Division, Bulletin 4, 1997. Available on-line at: <www.fao.org/docrep/W4347E/w4347e00.htm#Contents>.

6 J. A. Sarfoh, *Hydropower Development in West Africa: A Study in Resources Development* (American University Studies, Peter Lang Publishers, 1990).

7 G. Foley, W. Floor, G. Madon, E. M. Lawali, P. Montagne, and K. Tounao, *The Niger Household Energy Project: Promoting Rural Fuelwood Markets and Village Management of Natural Woodlands*. World Bank Technical Paper No. 362, Energy Series. The World Bank, Washington, 1997.

8 World Bank, *A Survey of the future role of Hydroelectric Power in 100 Developing Countries*. Energy Department paper No. 17, Washington, 1984.

9 A. Daniel, Government to Review NEPA tariff in Niger Republic. *The Nigerian Guardian Online* <www.ngrguardiannews.com>, September 18, 2001.

10 A. Daniel, Hostility Looms as Niger Plans to Dam River Niger. *The Nigerian Guardian Online* <www.ngrguardiannews.com>, December 12, 2002.

11 M. T. Coe and J. A. Foley, Human and Natural Impacts on the Water Resources of the Lake Chad Basin, *Journal of Geophysical Research D: Atmospheres* (2001): 106(4): 3349–3356.

12 A. Jauro, Executive Secretary Lake Chad Basin Commission, <www.oieau.fr/ciedd/contributions/atriob/contribution/cblt.htm>.

13 C. Benson and E. Clay, *The Impact of Drought on Sub-Saharan African Economies: A Preliminary Examination*. World Bank Technical Paper No. 401, 1998.

14 IRIN, <www.reliefweb.int/IRIN>, June 30, 2000.

15 Andrew Furber, *Fulani and Zarma tribes pushed into Fighting by Desertification? ICE Case Studies*, <www.american.edu/TED/ice/nighe.htm>, 1997.

16 A. Tarhule, Environment and Conflict in West Africa. In Max G. Manwaring (ed.), *Environmental Security and Global Stability: Problems and Responses* (Lexington Books, 2002), pp. 51–83.

17 United Nations Office for Emergency Operations in Africa (UNOEOA),

Special Report on the Emergency Situation in Africa 1986. *Review of 1985 and 1986 Emergency Needs*, New York.

18 R. H. Faulkingham, Subsistence strategies of Hausa farmers under drought conditions. In M. M. Horowitz (ed.), *Colloquium on the Effects of Drought on the Productive Strategies of Sudano–Sahelian Herdsmen and Farmers* (Binghamton, NY: Institute for Development Anthropology, 1976), pp. 32–42.

19 African Newswire Network (ANN), August 25, 2000 <www.africanewswire.com>.

20 African Newswire Network (ANN), August 17, 2000 <www.africanewswire.com>.

21 T. F. Homer-Dixon, On the Threshold: Environmental Changes as Causes of Acute Conflict, *International Security* 16:2 (1991): 77–116.

22 T. F. Homer-Dixon, Environmental Scarcities and Violent Conflict: Evidence from Cases, *International Security* 19:1 (1994): 5–40.

23 M. A. Novicki, Improving Africa's Water Supply: Special Initiative Seeks Greater Access for Farmers and the Poor, *Africa Recovery*, United Nations Publications, 2000.

24 F. A. Street-Perrott, J. A. Holmes, M. P. Waller, M. J. Allen, N. G. H. Barber, P. A. Fothergill, D. D. Harkness, M. Ivanovich, D. Kroon, and R. A. Perrott, Drought and dust deposition in the West African Sahel: a 5500-year record from Kajemarum Oasis, northeastern Nigeria, *The Holocene* 10(3) (2000): 293–302.

Part II

Management of Limited Resources

CHAPTER 6

Is Joint Management of Israeli–Palestinain Aquifers Still Viable?

ERAN FEITELSON

Israelis and Palestinians are two closely inter-related parties sharing the same aquifers. This is particularly true with regard to the Mountain Aquifer (see figure 6.1). This aquifer, composed of three sub-basins, supplies approximately a third of the Israeli water consumption,[1] and is the source of almost all the water supplied to the Palestinians in the West Bank. Due to the properties of this aquifer, it has long been suggested that it should be managed jointly.[2] However, there is only scant international experience in the management of transboundary aquifers. Hence, if the two parties do indeed intend to manage this shared resource judiciously, it is likely they will need to come up with innovative management structures. A series of such options were proposed in a cooperative study undertaken since 1993.[3] These are described in the following section. In practice a coordinated management structure was established in the interim agreements (Oslo B) signed in September 1995. This structure is composed of a joint water committee (JWC) and joint supervision and enforcement teams (JSETs). Early experience with this structure led to arguments that it is insufficient, and that there would be a need to move to more sophisticated structures. Practical steps to this end were also proposed in lieu of the permanent status negotiations that were expected at the time.[4]

Nevertheless, the JWC and JSET structure operated, though imperfectly, until the outbreak of violence in September 2000. As a result of the breakdown of the peace process at that point, the structure has in effect ceased to operate, though contacts between the parties on the water issues

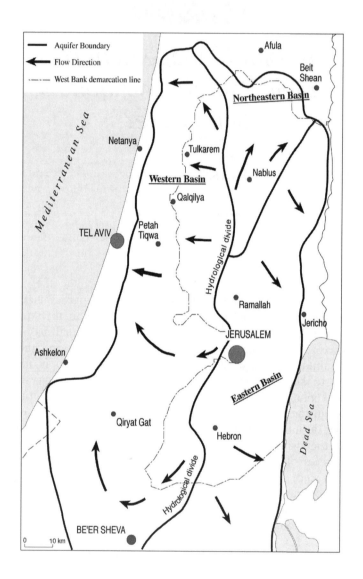

Figure 6.1 The Mountain Aquifer

continue. The question this essay asks is whether, given this setback, joint management structures between these two parties are still viable.

To address this question the chapter begins by briefly reviewing the set of options identified in prior work. Then, the implications of a complete breakdown in relations, resulting in separate management, are reviewed. A discussion of these implications shows that there are still options that may be worthwhile pursuing. Some steps for advancing such options are raised in the conclusion.

The Options: A Brief Review

There are four basic options to manage a shared aquifer: separately, in a coordinated manner, jointly, or by delegating responsibility to an outside body.[5] Under separate management each party sets its own policies, drills its own wells, determines its own extraction rates, and sets its own standards. Coordinated management implies that each party manages the aquifer within its own territory but coordinates its actions with the other party. This is in essence the type of structure established under the Oslo B agreement.[6] Joint management is the situation whereby a single institutional structure is established to carry out certain tasks viewed by the parties as the most crucial for adequate management of the aquifer. A fourth possibility is to delegate responsibility for the aquifer, or for some management tasks, to an external body. This could be a regional or international body, or a privately-owned corporation.

In practice, the sustainable management of any aquifer requires that many actions be undertaken.[7] These include determination of pumpage regimes and rates, drought policies, protection measures, monitoring, enforcement of restrictions on pumpage and land use, recharge enhancement projects, wastewater treatment standards and reuse policies, and crisis management measures. In addition, coordination of research as well as monitoring and sharing of data/models/expertise facilitate a more effective management regime. Thus, any structure for managing a shared aquifer can potentially address multiple issues. The extent to which it does so is a function of the terms of reference set for the structure. Therefore, there are many options for transboundary bodies, ranging from bodies that coordinate a single activity to bodies that manage the aquifer comprehensively.

In previous work, a flexible-sequential framework for the management of shared aquifers was proposed.[8] It suggests that initially a limited number of activities be undertaken jointly. These could serve as the basis for cooperative management structures. Additional activities would then be added to the purview of the structures over time. The added activities could lead to one of five basic orientations, according to the rationale chosen. Alternatively, from the second stage onward, activities could be added so as to expand the scope of the structure to include additional orientations. The five possible orientations identified were:

1. Resource protection, whereby the activities are geared towards the protection of the aquifer;
2. Crisis management, whereby the focus is on managing crisis situations, as in such situations the most contentious circumstances arise;
3. Efficient water use, whereby the focus of the structure is assuring the efficient use of water, most commonly through a trading system;
4. Effective water supply, where the management of the extraction of the water and its distribution to consumers, and perhaps also the collection and treatment of sewage, are entrusted to a third, private, company;
5. Comprehensive–integrative management, whereby an attempt is made from the outset to manage all aspects of the aquifer. This is the direction suggested in the Bellagio draft treaty.[9]

The choice of orientation is clearly a matter for negotiations. Still, recommendations were made regarding this choice.[10] In essence it was suggested that as resource protection and crisis management seem the most imminent concerns, it might be most appropriate to focus on them in the early stages. Establishing transboundary markets or franchises were seen as more problematic, given the complete inexperience with such structures. A comprehensive–integrative approach is also seen as an unlikely early structure, given the multiple facets that will need to be met, in a situation where government actions are increasingly questioned and scrutinized. From a theoretical vantage point it was suggested that cooperation begin with structures that imply relatively low costs, such as coordinated structures.[11] Yet, such transaction costs are likely to be affected by the overall relations between the parties. As these relations deteriorate the transaction costs are likely to escalate, thus reducing the likelihood of any cooperative arrangement.[12]

Issues Raised by Separate Management

If transaction costs of establishing cooperative management regimes escalate, the likelihood that the aquifers will ultimately be managed separately increases. If a separation regime can endure it is indeed unlikely that a cooperative structure will replace it, given the loss of confidence between the parties since September 2000, and hence the rising transaction costs. To assess whether there is a chance that a separate management regime will endure, it is necessary to identify and assess the issues such a regime raises. These are outlined in figure 6.2.

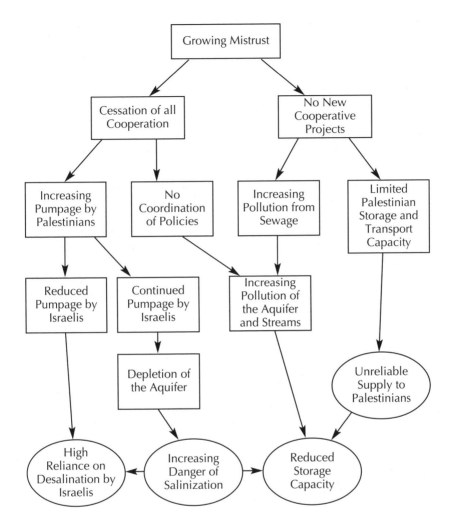

Figure 6.2   The Implication of Separation.

Under separate management each side will determine its own pumping regime. The Palestinians are likely in this case to increase their extractions, given the current shortages and their desire to reduce their dependence on Israel.[13] Israel can respond to this likely development in one of two ways. It can reduce its pumping so as to assure a sustainable yield, or continue to extract all it extracts currently. While the first option may lead to confidence building and thus possibly to better results in the long term,[14] it is counter-intuitive and contradicts much of the current Israeli thinking.[15] Hence, it is more likely that the result will be a race to the bottom, as the aquifer will increasingly be over-pumped. The full implications of such massive over-pumping are not fully known, due to the uncertainty regarding underground saline water bodies.[16] Still, it is clear that the threat of salination of the aquifer will rise, and that water levels will drop, consequently raising extraction costs.

A second issue that is likely to arise is that of pollution. This includes both pollution from landfills and other point sources, and wastewater collection, treatment and reuse. As most of the recharge zones are in the West Bank (see figure 6.1), it is likely that the Palestinians will have a disproportionate effect on this issue. The degree to which there will be point source pollution will be a function of solid and hazardous waste handling in the Palestinian entity and of land use regulations therein.[17] Given the multiple challenges that will face a new Palestinian entity, regardless of its exact boundaries, it is unlikely these issues will receive much attention, at least in the initial years. Thus, we can expect both point and nonpoint pollution to worsen as a function of population growth, economic growth, and changes in economic activities.

Currently, most of the wastewater generated in the West Bank flows untreated over the aquifer recharge areas. Some initial steps have been undertaken to build new wastewater collection and treatment plants on the West Bank with external funding, most notably the El Bireh wastewater treatment plant. However, much of the West Bank is not yet sewered, and almost none of the wastewater is treated to a secondary level.[18] Even with external funding it is likely that it will take considerable time before these problems are addressed. Moreover, the success of wastewater treatment is a function of the level of maintenance. Therefore, the mere construction of treatment plants does not assure that the wastewater will indeed be treated properly.

The wastewater treatment and reuse problem becomes even more difficult when the wastewater crosses boundaries. This is likely to occur in the Tulkarem, Qalqilya, and Jerusalem regions. In the Tulkarem region, a local level agreement between the municipality and the bordering Israeli regional council was reached. The question whether it will be implemented when the violence subsides is, however, moot at this point. In a study of the options for managing the sewage of the Jerusalem region, Feitelson and Abdul-Jaber show that separate management is the most costly, and least effective, option and suggest that third-party involvement by privatization should be explored.[19] If the levels of distrust preclude any cooperation it is quite obvious that pollution from raw sewage is likely to remain an unresolved problem.

Water supply to the Palestinians in the West Bank will continue to be a problematic issue if the systems are totally separated, especially in drought situations. Given the lack of storage capacity in the Palestinian territories, and the absence of conveyance systems in a north–south direction, except for the Israeli national water carrier, the Palestinians will find it difficult to balance the temporal and spatial variations in supply and demand. Hence, the Palestinians will face significant difficulties in assuring reliable water supply without Israeli assistance in conveyance and augmentation. Thus, even if Palestinians increase their extractions from the aquifers, the population, especially in the cities along the national water divide, may still suffer supply problems, especially in summer and drought years.

Discussion

If indeed the Mountain Aquifer will be managed separately, none of the issues noted earlier will likely be addressed, especially if any attempt to address them is perceived as providing the other side with a free rider option. Moreover, even if attempts are made to address some of these issues by any party (ignoring the free rider problem), the costs associated with such solutions will be considerably higher than in any cooperative mode.[20]

Cooperation between the two parties has two potential benefits. One, any cooperative agreement will impose external obligations on the two parties that may induce them to undertake actions that they may not do otherwise. For example, if the water for wastewater exchange idea[21] is adopted, Israel will be required to augment Palestinian supplies, particularly in drought situations,[22] and at the same time Palestinians will be obliged to treat their wastewater to a pre-specified level (probably secondary). None of this would occur in the absence of cooperation. Two, cooperation will allow for greater cost-effectiveness. It will allow for exploitation of economies of scale, better use of resources, and more effective data generation and use.

From any long-term perspective it is obvious that separation is an inferior option. This will remain true also after the introduction of large-scale desalination in Israel, as long as desalination will remain more costly than pumping from the aquifer. As long as this condition holds true the deterioration of the aquifer, and the subsequent rise in the cost of supplying clean potable water from it, will continue to imply an increase in the overall cost of water supply (as the substitution of desalinated water for groundwater will imply a higher cost).[23]

Over time the deleterious effects of separation are likely to become increasingly apparent. However, groundwater issues are generally less perceptible than other issues (including surface water issues), and the ability to rectify the damage to groundwater is limited and costly.[24] Thus, it is likely that by the time the damage is apparent enough to generate action it may be very late and much of the damage could be irreparable.

All of the adverse results of separation are well known to both parties, or at least to the experts on both sides. The problem in establishing cooperative regimes is largely an outcome of what Miriam Lowi termed "upper politics."[25] As she correctly pointed out, no agreement on water is likely unless it conforms to the outlines of interests framed by the upper politics. At the same time, all international negotiations are constrained by domestic politics, as all such agreements have to be ratified domestically.[26] In the upper politics field it is important to note that the parties did not return to the pre-Oslo no recognition stance. Rather, both sides state that eventually they would like to reach an agreement. Despite the violence, there are ongoing discussions between the parties on various issues. It is likely, therefore, that at some point the parties may search for issues where they can reach agreements that will be acceptable to their domestic constituencies, but will not seem trivial. These will have to be issues that do not compromise core beliefs, but are seen to provide tangible advantages. If this does indeed come to be, water can become a potential field for such agreements. The relative invisibility of water, and particularly ground-water, combined with the ability to generate tangible benefits, especially better and more reliable supply to the Palestinians, and less pollution to the coastal streams for the Israelis,[27] may help make water into a prospective confidence re-building measure. Moreover, if the confidence building measures within the water field, advanced by Haddad et al. prior to the breakdown in negotiations,[28] are implemented, the transaction costs of cooperation may be somewhat reduced again.

I have argued elsewhere that under the current circumstances the trans-action costs of the third-party involvement option may rise less than those of other options.[29] Yet, the implementation of this option raises a slew of issues. These include the questions of how differences will be adjudicated in a transboundary context, what will happen when certain customers default, what will be the legal basis for such structures in transboundary situations, how capital costs are to be recovered, and how the oversight of the franchisee will be structured.[30] Moreover, it is necessary to understand what is likely to happen when a two-party situation turns into a three-party situation, as will take place when an international firm assumes responsi-bility for any part of the shared water system. That is, the likelihood that Israelis and Palestinians will cooperate *vis-à-vis* the international firm (so as to get the best deal and level of services) has to be compared to the like-lihood that the international firm will play the two sides against each other thereby worsening the distrust among them.

To answer these questions it is necessary to conduct further research. Due to the inexperience with this type of structure in any similar context, such research is actually a requisite before this option can be seriously considered. However, under the current circumstances the ability to conduct cooperative research on such issues has been seriously compro-mised. Yet in the absence of such research, the ability to advance innovative structures that may allow the parties to avoid the separation default is limited, and hence the viability of such innovative cooperative options cannot be assessed yet.

Conclusions

The current trend in the management of the shared Israeli–Palestinian aquifers is towards separation. However, due to the close inter-relationship between the water sectors of the two parties, and the attributes of the aquifers, this option is probably the worst from a resource management perspective. In the long term it seems therefore that some form of cooperative management will be needed, as was recognized already in the Oslo B agreement. Yet by the time the damages of separate management will become apparent, much damage may be done to the detriment of both parties. Moreover, the greater the damage the greater the cooperation that will be needed to mitigate it. Actually, even the coordinated management structure established in the interim agreement was arguably insufficient.[31] Thus, the two parties will probably need to discuss again cooperation options, regardless of the current impasse.

The main obstacle to greater cooperation is the loss of trust between the two parties. This loss pertains both to the existing coordination structure and to the good intentions and commitment of the other party to peaceful resolution of differences. At the same time, if and when the two parties will seek an area where agreements are feasible, which can be acceptable to wide domestic audiences, water may stand out as such an area. Thus, it seems likely that negotiations over water issues will resume at some point. However, they will be overshadowed by the loss of trust.

To overcome this impediment it is suggested that the confidence building measures identified by Haddad et al. will be indeed implemented.[32] These include changes in the way the institutions function and an augmentation of Palestinian water supply. However, such steps are likely to be insufficient. In particular, there will be a need to seek concurrently the cooperative structures that may be most appropriate for the current situation and that may be conducive for further agreements. It is suggested that structures based on third-party involvement may be particularly suitable for this purpose. In addition to their purported effectiveness, benefits may turn a two-party situation, where the two parties are highly suspicious of each other, into a three-party game, where the two parties share interests *vis-à-vis* the third party. However, to implement this type of structure, substantial preparatory work is needed. Yet under the current circumstances, the capacity to undertake such work has been compromised. Thus, it seems there is a real need to advance such work, and foster cooperative research at present, even if the prospects for immediate implementation look remote.

Acknowledgments

This chapter is based in part on the work conducted in a long-term cooperative project on the joint management of the shared Israeli-Palestinian aquifers, funded by IDRC and the CRB Foundation, under the auspices of the Palestine Consultancy Group and The Truman Institute for Advancement of Peace. Professor Marwan Haddad, Taher Nasseredin, and Shaul Arlosoroff were co-researchers in this project, and have thus undoubtedly contributed significantly to this essay. However, the ideas expressed herein do not necessarily reflect those of any other participant in that project, or the agencies that funded or participated in it. They reflect only the views of the author, who also bears alone the full responsibility for any errors or omissions.

Notes

1   As the majority of settlements in the West Bank and Gaza are supplied through the Israeli system from pre-1967 Israel, Israeli consumption includes them.

2   E. Feitelson, Joint Management of Groundwater Resources: Its Need and Implications, in J. Cotran and C. Mallet (eds.), *The Arab–Israeli Agreements: Legal Perspectives* (London: Kluwer Law International, 1996).

3   E. Feitelson, Concept and Process for Mitigation of the Israel–Palestine Conflict Over Shared Groundwater Resources, in: G. J. Alearts, F. J. Hartveld and F-M. Patorni (eds.), *Water Sector Capacity Building: Concepts and Instruments* (Rotterdam: A.A. Balkema, 1999), pp. 185–94.

4   M. Haddad, E. Feitelson, S. Arlosoroff, and T. Nasseredin, Joint Management of Shared Aquifers: An Implementation Oriented Agenda, Final Report Phase II, Jerusalem: The Palestine Consultancy Group and the Harry S. Truman Institute for the Advancement of Peace, 1999.

5   M. Haddad, E. Feitelson, and S. Arlosoroff, The Management of Shared Aquifers, in: E. Feitelson and M. Haddad (eds.), *Management of Shared Groundwater Resources: The Israeli–Palestinian Case with an International Perspective* (Boston/Dordrecht/London: Kluwer, 2001), pp. 3–23.

6   M. Haddad, E. Feitelson, S. Arlosoroff, and T. Nasseredin, 2001, A Proposed Agenda for Joint Israeli–Palestinian management of Shared Groundwater, in: E. Feitelson and M. Haddad (eds.), *Management of Shared Groundwater Resources*, p. 476.

7   M. Haddad, E. Feitelson, and S. Arlosoroff, The Management of Shared Aquifers, in: E. Feitelson and M. Haddad (eds.), *Management of Shared Groundwater Resources.*

8   E. Feitelson and M. Haddad, A Stepwise Open-ended Approach to the Identification of Joint Management Structures for Shared Aquifers, *Water International* 23 (1988), pp. 227–37; E. Feitelson and M. Haddad, A Sequential Flexible Approach to the Management of Shared Aquifers, in: E. Feitelson and M. Haddad (eds.), *Management of Shared Groundwater Resources*, pp. 455–73.

9   R. Hayton and A. Utton, Transboundary Ground Water: The Bellagio Draft Treaty, *Natural Resource Journal* 29 (1989): 663–722.

10 M. Haddad, E. Feitelson, S. Arlosoroff, and T. Nasseredin, A Proposed

Agenda for Joint Israeli–Palestinian management of Shared Groundwater, in: E. Feitelson and M. Haddad (eds.), *Management of Shared Groundwater Resources.*

11   E. Feitelson, When Should Shared Aquifers be Managed Jointly?, paper presented at the National Groundwater Association annual conference, Las Vegas, December 13–14, 2000.

12   Ibid.

13   T. Nasseredin, Legal and Administrative Responsibility of Domestic Water Supply to the Palestinians, in: E. Feitelson and M. Haddad (eds.), *Management of Shared Groundwater Resources*, pp. 107–14.

14   G. Tal, Management of Shared Water Resources in a Situation Without Cooperation: The Mountain Aquifer Case, Unpublished seminar, Department of Geography, The Hebrew University of Jerusalem, 2001, in Hebrew.

15   The more common approach advocated within Israel is that Israel should continue to rely on the Mountain Aquifer plus seawater desalinization, while Palestinian supply will be augmented by seawater desalination, as seawater desalination is the only certain source. See, for a cogent example: D. Zaslavsky, *Sustainable Development of the Water Sector and the Fate of Agriculture* (Haifa: The Technion, 1999, in Hebrew).

16   Y. Harpaz, M. Haddad and S. Arlosoroff, Overview of the Mountain aquifer: A shared Israeli–Palestinian resource, in: E. Feitelson and M. Haddad (eds.), *Management of Shared Groundwater Resources*, pp. 43–56.

17   In Israel hazardous waste is disposed in a national site at Ramat Hovav in the Negev, while most solid waste is increasingly disposed in sanitary landfills that are not over the aquifer. Yet, this state of affairs is relatively recent, since the early nineties, and does not apply yet to all sources over the aquifer.

18   Palestinian Economic Council for Development and Reconstruction (PECDAR), Palestinian Water Strategic Planning Study (The Technical Assistance and Training Department, 2001), pp. 84–85; United Nations Environmental Programme (UNEP) Desk study on the environment in the occupied Palestinian territories, 2003.

19   E. Feitelson and Q. Abdul-Jaber, Prospects for Israeli–Palestinian cooperation in Wastewater Treatment and Re-use in the Jerusalem Region (The Jerusalem Institute for Israel Studies and the Palestinian Hydrology Group, Jerusalem, 1997).

20   The reason for the higher costs to each party is that separate water supply, sanitation and wastewater treatment (and re-use) systems will be built and maintained. Hence, economies of scale that could be achieved by combined systems will be forgone. An estimation of the extent of these losses will require a site-by site analysis, the assessment of which is beyond the scope of this chapter.

21   For an exposition of this idea, see: E. Feitelson, Water rights within a water cycle framework, in: E. Feitelson and M. Haddad (eds.) *Management of Shared Groundwater Resources*, pp. 395–405.

22   This has been shown to be the case in the Israeli–Jordanian case, as has recently been shown by Itay Fischhendler, Legal and Institutional Adaptation to Climate Uncertainty: A Study of International Rivers, unpublished manuscript, 2003.

23	Actually, the difference in social cost will be greater than implied here (in cost of supply), as desalinated seawater does not substitute for the ecological services provided by the springs and wetlands fed by the aquifers.

24	H. Gvirtzman, *Israel Water Resources* (Jerusalem: Yad Ben-Zvi, 2002, in Hebrew).

25	M. Lowi, *Water and Power: The Politics of Scarce Resource in the Jordan River Basin* (Cambridge: Cambridge University Press, 1993).

26	R. Putnam, Diplomacy and domestic politics: the logic of two-level games, *International Organization* 42, pp. 427–60.

27	See note 20.

28	M. Haddad, E. Feitelson, S. Arlosoroff, and T. Nasseredin, Joint Management of Shared Aquifers: An Implementation Oriented Agenda, Final Report Phase II, Jerusalem: The Palestine Consultancy Group and the Harry S. Truman Institute for the Advancement of Peace, 1999.

29	M. Haddad, E. Feitelson, S. Arlosoroff, and T. Nasseredin, A Proposed Agenda for Joint Israeli–Palestinian management of Shared Groundwater, in: E. Feitelson and M. Haddad (eds.), *Management of Shared Groundwater Resources.*

30	Ibid.

31	See M. Haddad, E. Feitelson, S. Arlosoroff, and T. Nasseredin, Joint Management of Shared Aquifers: An Implementation Oriented Agenda, Final Report Phase II, Jerusalem: The Palestine Consultancy Group and the Harry S. Truman Institute for the Advancement of Peace, 1999.

32	Ibid.

CHAPTER 7

The Southern West Bank Aquifer: Exploitation and Sustainability

DAVID J. SCARPA

The West Bank and Gaza may be considered as a land suffering stress. It is so defined under a simple mathematical equation of the total amount of water available per capita. A figure accepted by the World Health Organization is that a minimum per capita water requirement for survival is 100 m³/yr for domestic, urban, and industrial use plus a minimum of 25 m³/yr for fresh food for local consumption.[1] Globally, the per capita average is 800 m³/yr. Average per capita Palestinian provision is 82 m³/yr of which 26 m³ is for domestic consumption.[2]

It is apparent that water stress is not alleviated by scientific, technological, and engineering solutions alone. Allocation of water for agriculture in Palestinian areas derives from a complex mix of social and cultural pressures as well as economic and political considerations. Traditional farming is rain-fed, but the demands of food security and the need to provide livelihoods encourage the introduction of irrigated farming. These methods cannot be sustainable in this semi-arid region with its rapidly increasing population. It is possible that appropriately treated wastewater reuse could provide some irrigation water, but the water extracted from aquifers should be reserved for domestic use alone.

Water allocation in the West Bank historically has been determined to a large extent by the enormous water requirements of Israeli agriculture. Most of the increase in Israel's water use has been derived from the West Bank aquifers particularly during the 1970s and 1980s when they were entirely under Israeli control. The Israeli Civil Administration of Palestinian Affairs appears to have been dominated by powerful agricultural interests within Israel. Justification for retention of the occupied West Bank was seen by some Israeli politicians as essential for the water security of the State of Israel. The principal cause of low *per capita* water consumption and water shortage in the West Bank has been the artificial barriers placed on the Palestinian population during the Israeli occupation.[3] This prevented extended access to the water resources of the Mountain Aquifer. In the West Bank the political context has been the prime force governing water use and allocation, and therefore the patterns of degradation and depletion of groundwater resources. The occupation of the West Bank by the State of Israel from 1967 brought together by force two distinct and asymmetric social entities: the West Bank, an agrarian, capital-poor, low-income economy and the State of Israel, an industrial, capital-rich, high-income economy. The effect of the occupation was to warp the Palestinian social and economic structure in a number of ways, forcing the West Bank into a dependency relationship on the Israeli economy.[4] One of the guiding principles behind the organization of the water sector on the West Bank has been the integration of services into the Israeli network. The Israeli company Mekorot Water Company, Ltd., was able to accomplish this in 1982. Provision of such an essential resource as water gives the occupying power a sovereign status over the people thus occupied and dependent on this provision. This consequent limitation of Palestinian sovereignty over its people was recognized as the Palestinian Water Authority (PWA), which assumed responsibility for the provision of water to the Palestinian people only in name, but not in reality, even when it began drilling for water in the West Bank aquifers. The Israeli authorities routinely reject sites selected by the PWA and grant permits for alternative sites that they selected.[5]

This chapter deals with policies governing the aquifer development of the southern part of the eastern basin of the Mountain Aquifer over the last forty years.

BF	Beit Fajjar 3 (Mekorot)
H	Herodion wells 1-5 (Mekorot)
Jwc4	Jerusalem 4 (IDF Camp well)
Hb	Hebron 1 & 2 (Mekorot)
Hz	Hundaza (PWA)
PWA	Palestinian Water Authority wells 1,3 &11
Z	Zatara (PWA)
EH	Extra Herodion 1 & 2 (PWA)
BN	Bani Naim 2 & 3 (PWA)
Sh	Shdaimah (PWA)
Az	Azzariya (PWA)

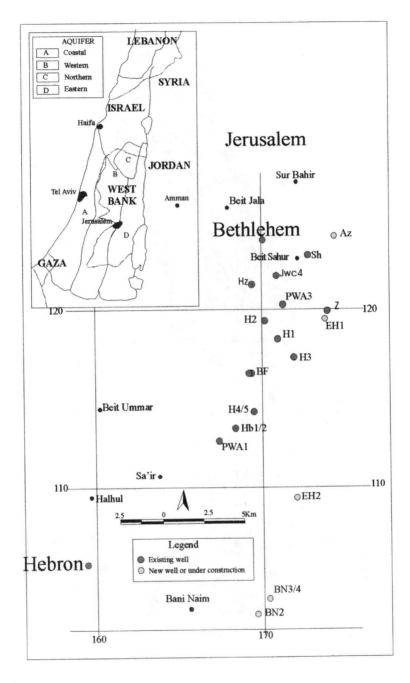

Figure 7.1 The Principal Production Water Wells in the Southern West Bank

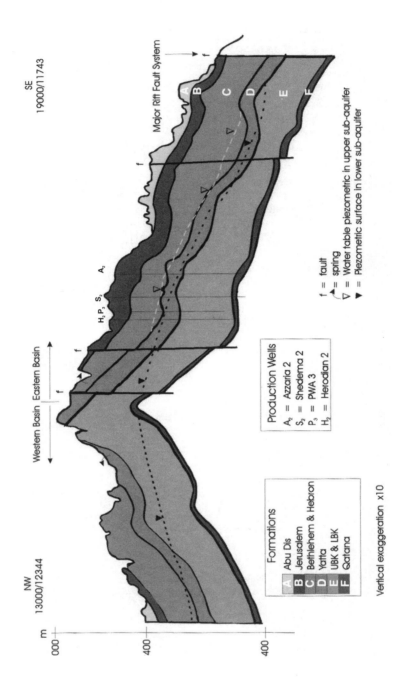

Figure 7.2

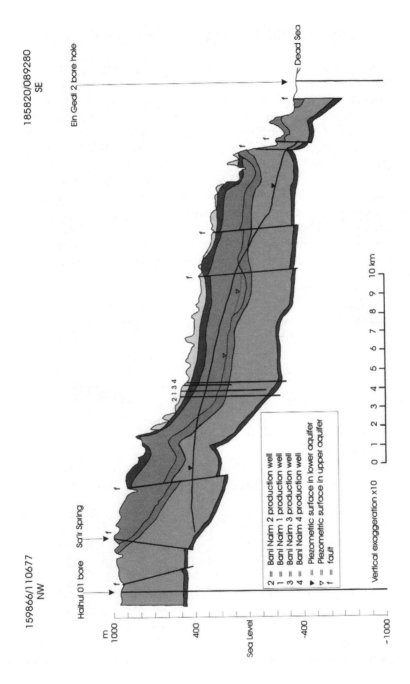

Figure 7.3

Water Wellfields in the Southeast West Bank

The principal source of drinking water in the southern West Bank is that part of the eastern basin of the Mountain Aquifer that drains to the Dead Sea (figure 7.1). The total thickness of this aquifer is between 800 m and 850 m in strata of Albian to Turonian age (107–89 million years BP), and is made up of two principal sub-aquifers; the upper, unconfined sub-aquifer, in Cenomanian to Turonian strata (94–89 million years BP), is between 50 m and 80 m higher than the lower, confined sub-aquifer, of Albian to Cenomanian age (107–90 million years BP). Borehole data indicate that the separation between these two sub-aquifers is the impermeable strata of bluish green clays and marls and some chalks in the Lower Cenomanian (figures 7.2 & 7.3).

Exploitation of this part of the aquifer has concentrated in the Herodian Beit Fajjar field and to a lesser extent further south in the Rihiya–Samu' wellfield, including the Fawwar wells. A more recent development is the Bani Naim wellfield.

Table 7.1   Production wells in the Herodian–Beit Fajjar Field [34]

Well	Coordinates	Elev. (mamsl)	Total depth	Aquifer	Swl (mbgl)	Yield* (m³/hr)	Pumpage (10⁶m³)
Beit Fajjar	16960/11510	728	307	Upper	159	230	1.72
Herodion 1	17092/11833	570	350	Upper	222	120	0.95
Herodion 2	17090/11933	563	770	Lower	258	336	2.80
Herodion 3	17085/11722	617	800	Lower	305	400	3.43
Herodion 4	16946/11408	685	691	Lower	326	249	1.82
Herodion 5	16946/11412	680	350	Upper	n/a	78	0.61
Hebron 1	16835/11310	710	705	Lower	320	30	n/a
Hebron 2	16835/11310	710	387	Upper	103	100	n/a
PWA3	17125/12025	610	740	Lower	320	250	1.8
PWA 11	16930/11635	752	752	Lower	380	250	1.8
Hundazza	16942/12137	625	672	Lower	325	350	2.0
PWA 1	16745/11245	745	600	Lower	360	250	1.8

Note: *pump-test rate, not sustainable yield or production rate.

The Herodian–Beit Fajjar wellfield[6] is located to the east of the line between Bethlehem and Hebron (table 7.1). Most of the production boreholes penetrate to depths of 350 m to 800 m, that is, into both the upper and lower sub-aquifers. Extraction is close to the area of recharge to obtain the greatest amount of high quality drinking water. The specific geology of the aquifer causes the flow direction in the aquifer towards the major discharge of the Feshkha springs along the Dead Sea shore. The water table is at about 450 m elevation above mean sea level (amsl) in the main recharge areas of the Hebron Mountains, and discharge at Feshkha is at about 410 m below sea level, that is, a head of 860 m over a lateral distance of about 20 km. In general terms, rainwater entering the recharge area of the aquifer in the Hebron Mountains will take about 40 years before it is discharged at Feshkha.[7]

History of the Development of the Southeastern Basins of the Mountain Aquifer

Exploitation of the southeastern parts of the Mountain Aquifer began under the Jordanian jurisdiction of the West Bank and was subsequently developed by Mekorot following the June 1967 war, and by the PWA following the 1995 Interim Accords.

1960–1967

The British consulting engineers, Rofe and Raffety,[8] contracted by the Hashemite Kingdom of Jordan, recognized the aquifer potential of what was to become the Herodion-Beit Fajjar well field. The consultancy team noted that population growth in the Bethlehem area from 1952–1961 was nearly static, while Jerusalem's growth during this time was 1.5%, considerably less than the national average for the kingdom. It was estimated that, at a rate of 1.5% per decade, the 1963 population of about 200,000 for the area including Ramallah, East Jerusalem, and Bethlehem, would rise to about 315,000 by the mid-1990s, or to 420,000 if the rate increased to 2.5% per decade. It was to satisfy the needs of this growing population that this part of the Mountain Aquifer was to be developed.[9]

   The Beit Fajjar well was drilled and successfully pump-tested, fitted with pumps and a distribution network, providing domestic water supply at a rate of 1,200 m³/day. This exploitation and allocation was clearly understood as a statement confirming Jordan's sovereignty over the West Bank. The development of this resource, based on carefully considered population projections and a sustainable exploitation of the field, could not be realized according to the Jordanian plan because of the Israeli occupation of the West Bank following the 1967 war. Further development by Israel in this part of the Mountain Aquifer ensued.

1967–1993

The Israeli company, TAHAL Consulting Engineers, Ltd., drilled the first two wells in the Herodion series in 1971. Herodion 3 was drilled in 1983 to the extraordinary depth of 800 m, into the confined sub-aquifer, with the help of Canadian money, on the condition that there be equitable supplies for both Palestinians and Israelis. This concession for the Palestinians was achieved by the skilful diplomacy of Bethlehem's Mayor Freij. It has, however, never been the case that an equitable *per capita* distribution was achieved. Israeli settlers receive an order of magnitude more water *per capita* from this part of the aquifer than the Palestinians (800 m^3 and 82 m^3 per capita respectively).[10]

The original Beit Fajjar well was twice replaced by deeper wells on the same site. The present well, drilled in 1988, has a total depth of 307 m. Herodion 4 was drilled into the confined sub-aquifer to a depth of 691 m in 1986 and Herodion 5 was drilled on the same site into the unconfined sub-aquifer in 1993 to a depth of 350 m.

These six wells have been lowering the static water table at an alarming rate. Table 7.2 illustrates that decline in three of the wells. The steady fall in the static water table is particularly evident in the drier years between 1979 and 1989 when a drop of about 30 m was recorded in Herodion 3. In 1987 the static water level in Herodion 4 stood at 377 m amsl, but had fallen to 346 m by the middle of 1991 and recovered to 359 m by 1995. The recovery simply reflects the effects of a twice the average recharge following the very heavy rainy season of 1991–92.[11]

Table 7.2   Lowering of static water levels in selected wells[13]

Well	Observation period	Drop in water table over period	Annual rate of decline in water table
Herodion 2	1975–1997	37.43 m	1.7 m
Herodion 3	1981–1997	85.33 m	5.33 m
Herodion 4	1986–1997	18.75 m	1.7 m

Thus, even before any of the new PWA wells were in production at the turn of the century, this part of the Mountain Aquifer was already being seriously mined by over-pumping.[12]

1993–2002

The excessive exploitation of the aquifers below the West Bank to satisfy settler and other Israeli demands at the expense of the local Palestinian population meant that the latter remained well below the minimum health requirement for water as defined by the World Health Organization.[13] It is within this context that water, as a fundamental political issue, touching on sovereignty, human rights and the interpretation of International Law, entered the arena of the peace negotiations, particularly the bilateral talks between Israelis and Palestinians.

The Bilateral Interim Agreement[14] meant that both the Israeli and Palestinian authorities were committed to drilling more wells with extraction rates of up to 250 m^3/hr to satisfy the domestic needs of the Palestinian populations in the Hebron and Bethlehem districts. The responsibility to provide adequate water supplies to the Palestinian communities prompted the Palestinian Authority to embark on a policy of aquifer extraction that would further endanger the sustainability of an already vulnerable resource. The foreign funding agencies and construction firms did nothing to discourage this policy.

The principal protocol for implementing an interim accord between Israel and the Palestinians, with respect to water allocation, is Article 40 of the Interim Agreement together with its appendices. A Joint Water Committee (JWC) was commissioned by this protocol to deal with all water and sewage related issues in the West Bank during the interim period. There was a general lack of satisfaction on both sides that the activities and procedures intended and anticipated by the scientists involved did not, for the most part, take place during the interim period (1995–99).

The Water Resources Program presented an overview of potential supplemental production wells in the West Bank to cover Stage Two water needs up to the target of 51.4 Mm^3/yr. Deep wells were drilled for the PWA with USAID money, with Jordanian engineering firms supplying the plant and technical knowledge. The consortia had to seek permission to drill in places that were deemed acceptable to the Israeli authorities. Many sites thought suitable by both Palestinian and foreign hydrologists, did not get the necessary permits from the Israelis.[15]

The Palestinian Development Plan (PDP) (1999) dedicated 48% of its total budget over the specified five-year period (1999–2003) to the realization of the infrastructure sector of the PDP. Palestinian infrastructure had suffered serious neglect during the years of Israeli occupation of the West Bank (1967–1999). It was clear that the Palestinian territories required massive fundamental investment and overhaul. Within the infrastructure sector, the PDP's first priority was given to water and wastewater projects, which were allocated 43.7% of that sector's total investment budget over the five years. The provision of good quality drinking water was the major priority within the West Bank.

Three production well sites were identified for immediate development, extracting 5–8 Mm^3/yr for the Hebron and Bethlehem districts. Permits were required from the Israeli authorities and mutual agreement about the ideal location for the sites was rare. Permits for four production wells in the Herodion well field were granted. Under the conditions of the Interim Accords (1995), Israel was to provide additional water to the Palestinian Authority. TAHAL, drilled Hebron 1 and 2 in 1996 to provide domestic water for the city of Hebron. Recommendations were also made for future production wells in the eastern basin of the Mountain

Aquifer to provide the 51.4 Mm³/yr as stipulated in Schedule 7.b(6) of
Article 40 of the Bilateral Interim Accords (1995). USAID funded a further group of wells in the southeastern basins of the Mountain Aquifer as
shown in table 7.3. A monitoring well was drilled near Bani Naim, 10 km
southeast of Hebron, an area believed to have great aquifer potential.[16]
The Water Resource Program then identified several production sites
with a capability of providing an additional 17 Mm³/yr for the Ramallah,
Bethlehem, and Hebron districts. This amount still falls short of the projected addition of 28 Mm³/yr, for the Bethlehem and Hebron districts
alone for the year 2000.

Table 7.3   Wells completed and under construction, 2002

Well	Coordinates	Elev. (mamsl)	Total depth (m)	Sub-aquifer	Capacity (m³/hr)
JWC	417087/12189	584	720	Lower	250
Azzariya 1*	17429/12403	529	800	Lower	250
Azzariya 2	17386/11988	591	790	Lower	250
Azzariya 3	17578/12852	455	850	Lower	250
Bani Naim 1	16899/10117	576	700	Lower	250
Bani Naim 2	16957/10243	520	700	Lower	250
Bani Naim 3	17032/10332	500	400	Upper	250
Bani Naim 4	17032/10332	500	750	Lower	250
E. Herodian 1	17370/11940	622	800	Lower	250
E. Herodian 2	17199/10908	633	700	Lower	250

Source: Aliewi and Jarrar (2000) see note 16.
Notes: * In June 2001 Azzariya 1 became inoperative as a result of a grouting error.
(mamsl = metres above mean sea level).

In the opinion of Aliewi and Jarrar, possibilities may exist for a more
sustainable exploitation of the well fields to the south and west of the
Herodion field, that is, in the Riheyeh and Samu' areas.[17]

The Rihiya Well Field

Four wells were completed in the 1960s in the Fawwar Refugee Camp
between Dura and Yatta, and reservoirs with a network system provided
domestic water for the towns and villages of the Hebron District. Fawwar
1 was rehabilitated by a grant from France in 1996 (table 7.4); the other
three wells became obsolete. Subsequently a pipeline was laid between the
well and the town of Dura.

The Rihiya or Zif 1 well is located 2 km north of the town of Yatta and
was drilled between 1984 and 1988. It supplies the city of Hebron, the town
of Yatta, and other surrounding towns and villages. Zif 1 has a 38 m draw-
down and a specific capacity of 21.5 m³/day In September 2000, the pump
in this well became inoperative and could not be repaired because of the

Israeli/Palestinian confrontations. An arrangement was eventually made with Mekorot and the PWA for the Herodian wells to supply Yatta and other nearby Palestinian communities with water.

Table 7.4   Rihiya wellfield characteristics

Well	Coords.	Elev	Total depth (m)	Aquifer	SWL	Yield (m³/hr)	Pumpage (x Mm³)
Zif 1	15720/09625	700	495	Upper*	317	34	0.30
Fawwar	115620/19815	731	100	Upper+		69	0.53

Source: WBWD (1995).
Notes:
mbgl = meters below ground level.
SWL = static water level in meters below ground level).
*Cenomanian, + Turonian.

Following discussions with the West Bank Water Department (WBWD) during 1997, additional wells in the Rihiya area were drilled but failed and were abandoned.

Conclusion

Israeli economic, political, and military power have defined the conditions of water scarcity for the Palestinians in the West Bank since 1967.[18] During the whole period of occupation, water supply and consumption remained relatively stable although there had been a considerable rise in population and some increase in incomes. A UN report observed that Israeli water policies in the Occupied Territories created a 'man-made' water crisis that undermined the living conditions and endangered the health of the Palestinian people.[19]

The Palestinian Authority recognized that one of its principal responsibilities was to alleviate the water stress suffered by the local Palestinian people. The Interim Accords seemed to allow for this provision. Foreign donors responded to this obvious need and further exploitation of the West Bank aquifers ensued. But there is a serious question whether the current exploitation of the southern part of the eastern basins of the Mountain Aquifer is sustainable.

The rapid response of the eastern basin of the Mountain Aquifer to the very high rainfall season of 1991–92 encouraged a higher extraction rate from the aquifer by the Israeli authorities.[20] The rainy seasons of 1999–2000 and 2000–2001 were marked by considerably below average rainfall and this was reflected in a lowering of the static water table. Once the new wells are in production, there is a strong probability that interference between so many wells, in such a relatively small wellfield, will result in a massive combined draw-down in the Herodion field.[21]

It must also be borne in mind that there is a natural lowering of the static water table in the unconfined aquifer as a result of the solution erosion effects of widening and deepening of joints and fissures produced in this karstic aquifer. This lowering of the water table is accelerated as the base level of groundwater drainage, the surface of the Dead Sea, drops at the dramatic rate of over 1 m/yr from 400 m to 412 m bmsl (1990–2001).

It is important to have accurate models that enable appropriate levels of exploitation within the particular circumstances of aquifer use. The needs of the groundwater stakeholders must be borne in mind by the management authorities, as well as an appreciation of the special social, political and economic situation at any one time. Sometimes, temporary overexploitation may be necessary before a sustainable development is achieved.[22]

The Israeli model of the eastern basin of the Mountain Aquifer[23] and the Palestinian model[24] are based primarily on the logs from the production and exploration wells. There are geophysical and radar *cave* technologies that could have been employed with capabilities to provide a much more accurate model of this karstic aquifer, but these were not used. The arbitrary delineation of the southern boundary in the Guttman–Zuckerman model,[25] coupled with remarks in Aliewi and Jarrar[26] suggesting that the southernmost parts of the aquifer could be the most promising, raise doubts as to the scientifically disinterested status of these models.

It is the opinion of some hydrologists and engineers[27] that the amount of aquifer water available in the eastern basin of the Mountain Aquifer, published in Article 40 of the Oslo 2 Accords (1995),[28] is an overestimation, based in part, on a misinterpretation of the nature of the Feshkha discharge. Only a relatively small amount of the water discharged at Feshkha represents recharge water into those parts of the aquifer being exploited by the recently drilled wells. The Feshkha springs and seepages, in fact, drain a very wide area and include 'fossil water' from the Nubian Sandstone aquifer.[29]

Could wise aquifer management, shared between Israeli and Palestinian authorities, save the aquifer, even in the foreseeable future? Although shared management of the Mountain aquifer has come under detailed study by Israeli and Palestinian academic and professional scientists and sociologists, there is some reluctance on the part of most politicians from both sides to endorse publicly their recommendations. The Joint workshops organized by the Truman Institute of the Hebrew University of Jerusalem and the Palestinian Consultancy Group have been taking place since the start of the Peace Process.[30] Such conferences, workshops, and joint publications seem to make little or no impact on decision or policymakers. Most of the shared aquifers of Israel and the Palestinian Authority will be able to supply water for domestic consumption for less than a quarter of a century. Feitelson and Haddad[31] assume that *per capita* water allocation for domestic purposes will be equal for both Israelis and Palestinians. These authors go on to point out that the negotiators striving to achieve this equitable supply must take into account climate change and the consequent hydrologic and water quality fluctuations and not simply concentrate on amounts of water currently available.

If it is the case that Palestinian political discourse will engage in a consideration of a solution to the present water crisis by means of an increased importation of food, it thereby engages in global political considerations. The validity of the *virtual water* solution[32] is dependent on cheap food being available on the world market and available to relatively poor economies like those in the West Bank and Gaza. The exporting countries, where there is adequate round-the-year soil moisture, may subsidize this food production. The West Bank and Gaza would therefore have to diversify its economy away from irrigated agriculture towards industry, particularly service industry, that provides much better returns for the water resources, making it possible to import the bulk of its foodstuffs.

Absence of cooperation in water is an historical fact for Israelis and Palestinians. It may be said that the tragedy of the commons in the Mountain Aquifer system has been avoided up to now by Israel's control as occupier. The Palestinians are still not in control of their principal source of fresh water. The key issues for Palestinians, in the view of Gass et al.,[32] do not center on a threat to future water supply but on access, control, and justice. However the threat remains. It is not, at present, within the capacity or the will of the Palestinian Authority to make the necessary adaptations to an economy, at present extremely fragile, that could remove that threat. The sustainability of the eastern basin of the Mountain Aquifer is dependent, as so much else, on a valid peace process, leading to freedom and security for both Israelis and Palestinians. If the way forward is shared management of the Mountain Aquifer, then a level of respect and trust, allowing a mutual co-operation in trade, investment, and infrastructure, is essential, but this is not apparent as a realizable possibility in the near future. Will there be sufficient time to save the aquifer?

Notes

1   World Health Organization (WHO), *Drinking Water Guidelines* (New York, 1995).

2   Palestinian Central Bureau of Statistics (PCBS), *The Demographic Survey in the West Bank and Gaza Strip, Final Report* (Ramallah, 1997); Palestinian Academic Society for the Study of International Affairs, *PASSIA Diary 2001* (Jerusalem: PASSIA, 2000), p. 261.

3   G. Gass, A. MacKenzie, Y. Nasser, and David Grey, *Groundwater Resources Degradation in the West Bank: Socio-economic Impacts and Their Mitigation,* unpublished report for PWA and funding agencies (1st draft), undated, pp. 15–16.

4   Ibid., p. 12.

5   Camp Dresser and McKee International Inc. (CDM) for United States Agency for International Development. *Two-Stage Well Development Study for Additional Supplies in the West Bank, Task 19, Stage 2, Well Development Study,* 1997.

6   Alfred Abed Rabbo, David J. Scarpa, and Ziad Qannam, A Study of the Water Quality and Hydrochemistry of the Herodion–Beit Fajjar Wells, *Bethlehem University Journal* 17 (1998): 10–28.

7   David J. Scarpa, Eastward groundwater flow from the Mountain Aquifer, in Isaac Jad and H. Shuval Hille, *Water for Peace in the Middle East* (Amsterdam: Elsevier Science Publications, 1994), pp. 193–203.

8   Rofe and Raffety, Consulting Engineers, *Jerusalem and District Water Supply, Geological and Hydrological Report* (London: Ministry of Water and Irrigation, The Hashemite Kingdom of Jordan, 1963).

9   Alfred Abed Rabbo, David J. Scarpa, and Ziad Qannam, A Study of the Water Quality and Hydrochemistry of the Herodion–Beit Fajjar Wells, *Bethlehem University Journal* 17 (1998): 10–28.

10   Palestinian Academic Society for the Study of International Affairs, *PASSIA Diary 2001* (Jerusalem: PASSIA, 2000), p. 274.

11   Y. Guttman and C. Zukerman, *A Model of the Flow in the Eastern Basin of the Mountains of Judea and Samaria from the Far'ah Stream to the Judean Desert,* Washington: Water Division, the Hydrology Department, Water Planning for Israel, Inc (TAHAL) in Hebrew, 1995 [trans. Daniel Pedersen, 1996]; Amjad Aliewi and Aiman Jarrar, *Technical Assessment of the Potentiality of the Herodian Wellfield Against Additional Well Development Programmes* (Ramallah: Palestinian Water Authority, 2000).

12   Ibid.

13   World Health Organisation, *Drinking Water Guidelines* (New York: WHO, 1995).

14   Israel–Palestinian Bilateral Negotiating Team, *Water Supply and Sewage Disposal,* Washington: Interim Accords, Article 40, 1995.

15   Camp Dresser and McKee International Inc. (CDM) for United States Agency for International Development. *Two-Stage Well Development Study for Additional Supplies in the West Bank, Task 19, Stage 2, Well Development Study,* 1997.

16   Amjad Aliewi and Aiman Jarrar, *Technical Assessment of the Potentiality of the Herodian Wellfield Against Additional Well Development Programmes* (Ramallah: Palestinian Water Authority, 2000).

17   Ibid.

18   G. Gass, A. MacKenzie, Y. Nasser, and David Grey, *Groundwater Resources Degradation in the West Bank: Socio-economic Impacts and their Mitigation,* unpublished report for PWA and funding agencies (1st draft) undated, p. 12.

19   United Nations, *Water Resources of the Occupied Palestinian Territory,* United Nations (Washington DC, 1992).

20 Tony Allan, *The Middle East Water Question: Hydropolitics and the Global Economy* (London: I.B. Tauris, 2001).

21 Amjad Aliewi and Aiman Jarrar, *Technical Assessment of the Potentiality of the Herodian Wellfield Against Additional Well Development Programmes* (Ramallah: Palestinian Water Authority, 2000), p. 10.

22 Emilio Custodio, *The Complex Concept of Overexploited Aquifer*, Papeles del Proyecto Aguas Subterraneas, 2nd ed. (Madrid: Fundacion Marcelino Botin, 2000), p. 5.

23 Y. Guttman and C. Zukerman, *A Model of the Flow in the Eastern Basin of the Mountains of Judea and Samaria from the Far'ah Stream to the Judean Desert*, Washington: Water Division, the Hydrology Department, Water Planning for Israel, Inc. (Tahal) in Hebrew, 1995.

24 Amjad Aliewi and Aiman Jarrar, *Technical Assessment of the Potentiality of the Herodian Wellfield Against Additional Well Development Programmes* (Ramallah: Palestinian Water Authority, 2000), p. 10.

25 Palestinian Academic Society for the Study of International Affairs, *PASSIA Diary 2001* (Jerusalem: PASSIA, 2000), p. 274.

26 Ibid., pp. 8, 10

27 Gert De Bruijine, Jenifer Moorehead, and Wa'el Odeh, *Water for Palestine: a Critical Assessment of the European Investment Bank's Lending Strategy in the Rehabilitation of Water Resources in the Southern West Bank* (Brussels: Reform the World Bank Campaign Report, Palestinian Hydrology Group, 2000); Amjad Aliewi and Ayman Jarrar, *Technical Assessment of the Potentiality of the Herodian Wellfield Against Additional Well Development Programmes* (Ramallah: Palestinian Water Authority, 2000), p. 10.

28 Israel–Palestinian Bilateral Negotiating Team, *Water Supply and Sewage Disposal*, Washington: Interim Accords, Article 40, 1995.

29 David J. Scarpa, West Bank Aquifers, *The Washington Report on Middle East Affairs* (2001): Vol. XIX, No. 9, p. 4.

30 Eran Feitelson and Marwan Haddad (eds.), *Joint Management of Shared Aquifers* (Jerusalem: Truman Institute and Palestinian Consultancy Group, 1995, 1996, 1997, 1998); Eran Feitelson and Marwan Haddad, *Management of Shared Groundwater Resources: The Israeli–Palestinian case with an International Perspective* (Dordrecht: Kluwer Academic Publishers, 2001).

31 Eran Feitelson and Marwan Haddad, *Management of Shared Groundwater Resources: The Israeli–Palestinian case with an International Perspective* (Dordrecht: Kluwer Academic Publishers, 2001), p. 483.

32 J. Anthony Allan, 'Virtual Water': a long-term solution for water short Middle Eastern economies? Proceedings of the 1997 Leeds Conference, Leeds: The British Association; *The Middle East Water Question: Hydropolitics and the Global Economy* (London: I.B. Tauris, 2001).

33 G. Gass, A. MacKenzie, Y. Nasser, and David Grey, *Groundwater Resources Degradation in the West Bank: Socio-economic Impacts and their Mitigation*, unpublished report for PWA and funding agencies (1st draft) undated, p. 12.

34 CDM (1997).

CHAPTER 8

Groundwater Salinization in the Jordan Valley – Quo Vadis?

AKIVA FLEXER, YAAKOV ANKER, LEA DAVIDSON,
ELIYAHU ROSENTHAL, ANNAT YELLIN-DROR, AND
JOSEPH GUTMAN

The objective of this study is to formulate a regional, operational water-management plan, whose goal is to minimize the deterioration and loss of high-quality ground water. In order to achieve this goal, it has been necessary to define and quantify the recharge origins in an improved manner and to fill the gaps in knowledge that have prevented the full understanding of groundwater salinization processes. The practical goal has been to produce a 3-D hydrogeological and hydrogeochemical model that defines the main factors and processes active in the area. This model contains all the relevant information for the formulation of a regional management plan. This project was based on the cooperation of multilateral research teams, consisting of Germans, Israelis, Jordanians, and Palestinians, that studied sustainable utilization of aquifer systems. It was supported by the German and Israeli Ministries of Science.

A few preliminary remarks should help define the problem. The multi-aquifer system of the Jordan Valley constitutes part of the Quaternary Dead Sea Group. It consists mostly of quaternary clastic sediments that create a multi-aquifer system. The possible sources of replenishment to the Jordan Valley multi-aquifer system are waters originating in adjacent aquifers on both sides of the Jordan Valley. The eastern, i.e. the Jordanian side, exposes sandy aquifers of Triassic and Jurassic age alongside dolomite and limestone aquifers of Cretaceous age.[1] The western, i.e. the Palestinian–Israeli side, discloses Cretaceous and Eocene dolomite and limestone aquifers.[2] The salinity of the Jordan Valley groundwater is high in comparison with that of its sources of replenishment. This implies groundwater salinization processes occur mostly in the contact zone between the source aquifers and the Jordan Valley multi-aquifer system but also within the clastic sequence forming the aquifer system. These salinization processes cause the deterioration of large amounts of high-quality groundwater.

General Geologic Setting

The southern Levant area (namely Israel, Jordan, and the West Bank and Gaza) is subdivided into 4 longitudinal N–S strips from west to east (figure 8.1):

1. The Coastal Plain, extending from the border with Lebanon in the north to the Egyptian border in the south and including the Gaza strip. The mountain foothills bound the Coastal Plain to the east and the Mediterranean Sea to the west.
2. The Mountainous Backbone: the Negev, Hebron, Jerusalem-Nablus, and Galilee.
3. The Dead Sea–Jordan Rift Valley. This is a major dislocation line that extends from the Red Sea (Gulf of Aqaba) northwards via the Beqa'a of Lebanon and terminates in the Taurus Mountains of Turkey.
4. The high mountains and elevated plateau of Jordan in the east.

Paleogeographic reconstruction[3] shows that the Mountainous Backbone initiated its uplift prior to the mid-Miocene (23–15 million years ago, MYA). The Dead Sea–Jordan Rift Valley, a remarkable plate boundary, started its activity during the mid-Miocene (15–10 MYA) following the opening of the Red Sea and the movement of the Arabian Plate northwards. Its typical tectonic habitat of horizontal left-lateral displacement was accompanied intermittently by vertical movements and the formation of pull-apart or extensional basins known by the term rhomb-shaped grabens.[4] These extensional opening movements were mostly active during the Late Miocene–Early Pliocene (7–4 MYA) and a chain of basins was clearly shaped. The most conspicuous basins from south to north are the Dead Sea, the Bet Shean Valley, the Sea of Galilee, and the Hula Valley. Successive marine ingressions advanced inland during the Late Miocene–Early Pliocene from the Mediterranean Sea. The sea water covered the Coastal Plain area and reached the north–south mountainous backbone of cis-Jordan. East–west oriented topographic depressions enabled sea water to invade the Jordan Valley. Evaporation of the water caused deposition of gypsum, halite, and post-halite minerals in the Jordan Valley.[5]

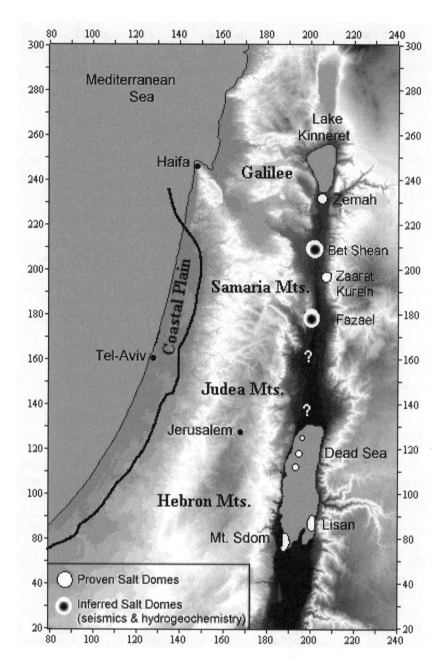

Figure 8.1 Location map showing the salt bodies along the Dead Sea–Jordan Rift Valley and the eastern boundary (heavy solid line) of the Mavqiim Evaporitic Formation in the Coastal Plain (modified after Flexer and colleagues[6]).

Geohydrology

The groundwater in the Jordan Valley (figure 8.1) flows through the porous permeable rocks of the Quaternary Dead Sea Group (formed 2 MYA) forming a multi-aquifer system. The major sources of recharge to the Jordan Valley aquifer system are located in the high mountains bordering the valley from both the western and eastern sides.

The West Side Story

The mountain aquifer or the Judea Group aquifer (Cenomanian–Turonian, 100–90 MYA), exposed in the Jerusalem–Nablus mountains (figure 8.1), is a geologic succession composed of a thick (600-800 m) sequence of hard, karstic, and permeable limestone and of dolomite interbedded with shale beds of lower permeabilities (figure 8.2, overleaf). The mountainous range represents a major NNE-directed chain of folded beds (anticlines and synclines). The subsurface groundwater divide dissects the aquifer into an eastern and a western aquifer. Groundwater penetrating the eastern aquifer flows eastwards, and there is a risk of salinization at certain places where it meets evaporite (mostly salt) bodies.

Guttman and Rosenthal[7] and Guttman[8] pointed out that the Judea Group sequence in the eastern slopes of the Judea and Samaria Mountains contains two subaquifers, which are the major sources of fresh water to the Jordan Valley. The hydraulic gradient is from 450 m in the west to –350 m in the east. The flow characteristics in these subaquifers are not uniform but depend on local tectonic factors. In most of the aquifer, the chloride concentration is 50–100 mg L^{-1}.[9] The recharge sources to this aquifer are rain and spring-originated runoff that infiltrates into the aquifer as well as interaquifer flow of groundwater originating from the Judea Group aquifer (in the Beqaot–Argaman region). Interaquifer flow from the Judea to the Avedat aquifer is possible because of faulting and extreme variability in the thickness of the separating Mt. Scopus aquiclude (figure 8.2).[10] The groundwater flow patterns to the multi-aquifer system from the eastern aquifers require further study.

The Tale of the Eastern Side

The eastern Jordanian side exposes several geologic and hydrologic levels. Bender[11] describes the hydrological conditions and claims that Triassic and Jurassic sandstones underlie the Cretaceous calcareous and sandy sequences in North Jordan. The depth of these aquifers is rather high, more than 400 m in all places of interest for groundwater utilization. This system does not generally provide adequate conditions for economic utilization of

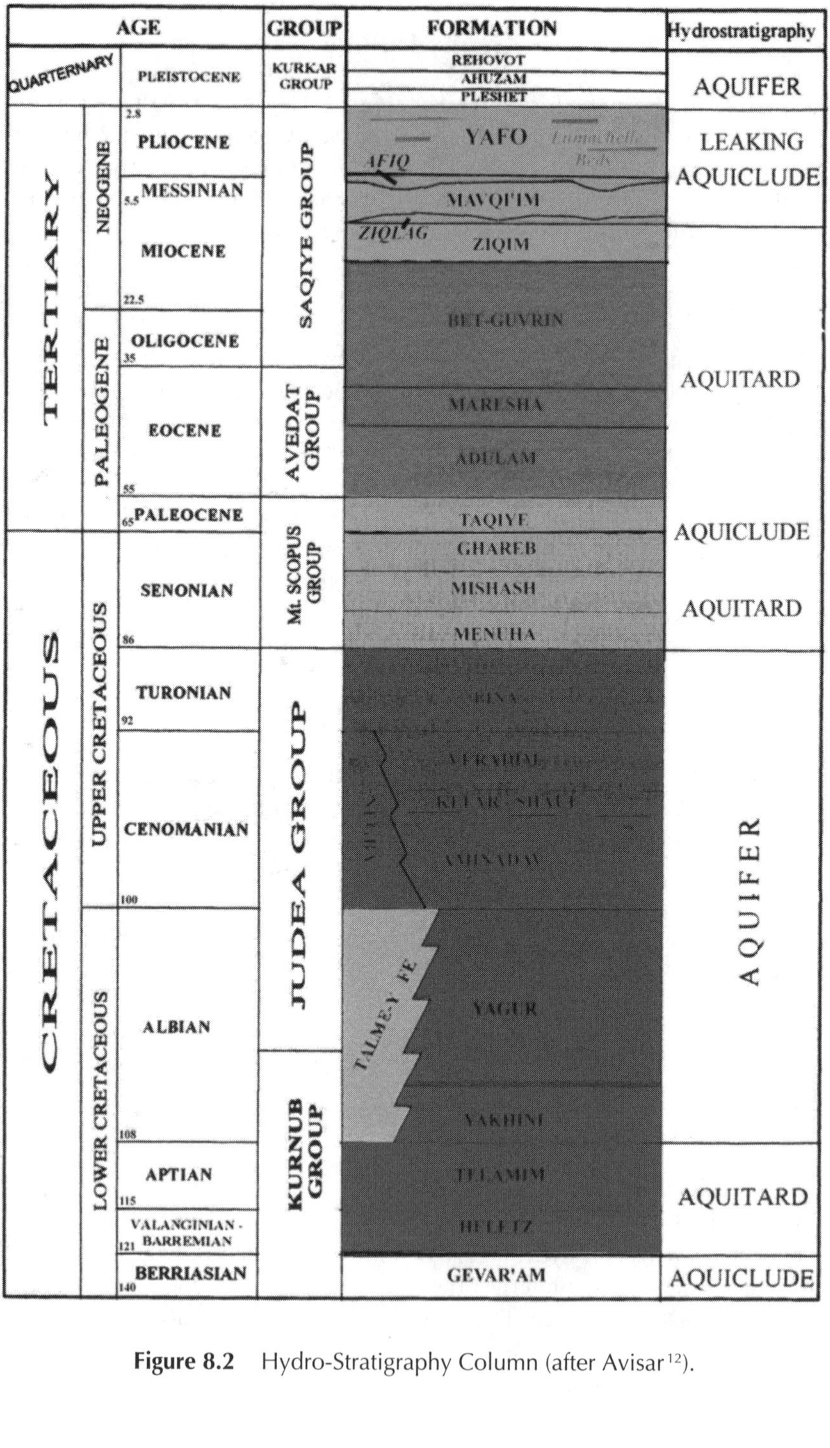

AGE			GROUP	FORMATION	Hydrostratigraphy
QUARTERNARY		PLEISTOCENE	KURKAR GROUP	REHOVOT / AHUZAM / PLESHET	AQUIFER
TERTIARY	NEOGENE	PLIOCENE 2.8	SAQIYE GROUP	YAFO *Lumachelle Beds* AFIQ	LEAKING AQUICLUDE
		MESSINIAN 5.5		MAVQI'IM ZIQLAG	
		MIOCENE 22.5		ZIQIM	
	PALEOGENE	OLIGOCENE 35		BET-GUVRIN	AQUITARD
		EOCENE 55	AVEDAT GROUP	MARESHA / ADULAM	
		PALEOCENE 65	Mt. SCOPUS GROUP	TAQIYE	AQUICLUDE
CRETACEOUS	UPPER CRETACEOUS	SENONIAN 86		GHAREB / MISHASH / MENUHA	AQUITARD
		TURONIAN 92	JUDEA GROUP	BINA	AQUIFER
		CENOMANIAN 100		VERADIM / KEFAR SHAUL / AMINADAV	
	LOWER CRETACEOUS	ALBIAN 108		TALME-Y FE / YAGUR	
			KURNUB GROUP	YAKHINI	
		APTIAN 115		TELAMIM	AQUITARD
		VALANGINIAN - BARREMIAN 121		HELETZ	
		BERRIASIAN 140		GEVAR'AM	AQUICLUDE

Figure 8.2 Hydro-Stratigraphy Column (after Avisar[12]).

groundwater owing to considerable drilling depths, generally low permeability, and poor chemical water quality. The main economic aquifers locally are the Cenomanian and Turonian limestone and Senonian brecciated flint beds, where they are directly recharged from rainfalls on outcrops along the mountain ridge.

Salinization Processes

The recharge to the Jordan Valley multi-aquifer system flows from the west, and probably also from the east, and consists of low-salinity water. However most of the groundwater in the Jordan Valley multi-aquifer system is brackish. Previous studies[13] have suggested that fresh groundwater flowing across the Dead Sea–Jordan Rift border faults to the Jordan Valley aquifer system becomes saline mostly due to relatively high concentrations of chloride and sulfate minerals found in the Jordan Valley itself. Likewise, unique ionic ratios (such as Mg:Ca > 1 and Mg:Cl ≈ 0.6) distinguish the groundwater of the aquifer complex from the water replenishing it.[14] Guttman and Rosenthal[15] and Guttman[16] noticed this phenomenon in the groundwater of the Jericho–Fazael region in boreholes near the border faults.

It seems that the unique geochemical character of the water in the Jordan Valley aquifer complex is a result of several geochemical processes. Z. B. Begin and colleagues[17] suggested that, in addition to the dissolution of gypsum layers in the Lisan Formation of the Dead Sea group, the major process is the mixing of fresh water with residual brines from the ancient Dead Sea. Yechieli and colleagues[18] suggested that fresh water infiltrates into the aquifer, leaching salts from the ground surface. The salts were probably deposited in soil crusts following runoff evaporation.[19] Additional enrichment in Mg, K, and SO$_4$ occurs during the percolation of the resultant solute to the aquifer. Final mixing of the solute with relatively fresh water occurs as it percolates into the aquifer. A similar process had been described for brackish groundwater in arid regions of Mexico and the United States.[20]

A different approach proposes that the salinization process could be caused by thermal, pressurized Ca-chloride brine injected into the aquifer.[21] It seems that the mixing of fresh water with Ca-chloride brines occurs in the region of border faults (mostly near major faults) and deteriorates the quality of the groundwater. Some observations indicate that the deterioration is accelerated by overpumping. The geochemical character of the saline component of the groundwater east to the Jordan River is considerably different from the Ca-chloride brines abundant in the west; these have yet to be investigated.

Another mechanism discussed by Flexer and colleagues[22] might be

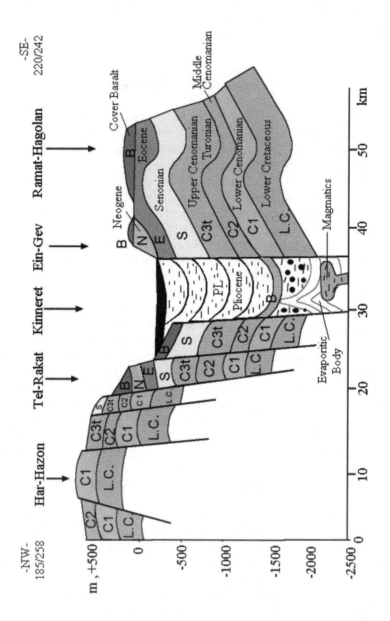

Figure 8.3 NW–SE geological cross-section through Lake Kinneret showing the possible hydrologic connection between the Cretaceous aquifers and the evaporitic body (after Flexer and colleagues[23]).

applicable for salinization of fresh water along the Jordan Rift Valley. They found that sporadic samples from the thick salt sequence found within the borehole wildcat Zemah-1 revealed the presence of evaporite minerals beyond the level of halite precipitation. The presence of post-halite minerals extending up to, and including, bischofite could not be excluded. The finding of more soluble K and Mg salts showed evidence of an advanced degree of seawater evaporation that acted, presumably cyclically, on seawater that intruded during Miocene–Pliocene times into the Rift Valley. Terminal evaporation of these brines resulted in the deposition of evaporitic (salt) bodies.

Figure 8.1 illustrates the distribution of evaporitic bodies along the Dead Sea–Jordan Rift. Fresh water percolating into the valley fill (figure 8.3) encounters and leaches evaporitic bodies, which are partially responsible for brines generated in the Rift Valley. Understanding these natural processes might contribute to the optimization and rationalization of groundwater exploitation and management.

Regional Operational Plan

The Zemah salt body and other evaporitic deposits may offer the key to understanding the salinization processes affecting the whole Dead Sea–Jordan Rift Valley. Identification of the evaporitic bodies as the primary brine source indicates a method for preserving water quality of the valley. If the leaching of these salts by inflowing fresh recharge water can be diminished, the resultant brine formation will be reduced. Therefore, we suggest controlling the amount of deep drainage of fresh recharge water from the surrounding areas to prevent it from leaching the evaporitic bodies and subsequently moving upward.

The three dimensional surface and subsurface geological and hydrological regimes in the area argue for the following:

1. Planning of a pattern of deep drill holes that will penetrate the main aquifers and the pumping of fresh water before it contacts the evaporitic salt bodies. These drill holes should be located mainly on the borders or the margins of the Jordan Valley.
2. Building a system of dams or barriers in the main wadis (stream beds) that could block the floods from the mountainous ridges on both sides of the Dead Sea–Jordan Rift Valley. This system would cause artificial percolation or recharge area of rain water directly into the aquifers.
3. Compiling the three-dimensional geological subsurface anatomy of the Dead Sea–Jordan Rift Valley by analyzing old and newly available seismic lines. Preliminary research shows the location of shallow

aquifers in the valley in areas free from evaporites. It is worth noting that even newly discovered brackish or saline ground water can be of use after desalination.

Conclusions

Geological, geophysical, and hydrological studies have delineated a three-dimensional picture of the water-rock relationships in the Dead Sea–Jordan Rift Valley. Our suggestion is to establish a drill-hole system that would pump fresh water beyond the contact of salt beds in the Rift Valley. A pattern of dams along the main wadis would increase fresh water percolation to the aquifers.

Notes

The authors can be contacted as follows: Yaakov Anker, Lea Davidson, Eliyahu Rosenthal and Annat Yellin-Dror, Department of Geophysics and Planetary Sciences, Tel Aviv University, Tel Aviv 69978, Israel; Yossi Gutman, Mekorot Water Company, 9 Lincoln St., Tel Aviv 61201, Israel.

1   F. Bender, *Geology of Jordan* (Berlin: Gebrueder Borntaeger, 1974), pp. 182–86.
2   Y. Guttman, Salinization along the western border of the Jordan Valley and the Dead Sea. In: *Proc. of the 13th GIF meeting on the Dead Sea Rift as a unique global site* (1997); Y. Guttman and E. Rosenthal, Mitzpe Jericho region – research of the salinization mechanism and water potential (Tel Aviv: Tahal, 1991), Report 01/91/46; Y. Guttman and H. Zuckerman, The Eastern aquifer in the Judea–Samaria region – calibration of numerical model (Tel Aviv: Tahal, 1995), Report 01/95/66.
3   Z. B. Begin and E. Zilberman, *Main stages and rate of relief development in Israel* (Jerusalem: Geological Survey of Israel, 1997), 63 pp. [in Hebrew, English abstract].
4   Z. Garfunkel and Y. Gat, The history and formation of the Dead Sea Basin in Niemi. In: *The Dead Sea – The Lake and its Setting*, edited by Z. Ben Avraham and Y. Gat (Oxford University Press, 1997), pp. 36–56.
5   Z. Garfunkel and Y. Gat, The history and formation of the Dead Sea Basin in Niemi. In: *The Dead Sea – The Lake and its Setting*, edited by Z. Ben Avraham and Y. Gat (Oxford University Press, 1997), pp. 36–56; I. Zak, Brines of the Dead Sea Basin: genetic aspects. In: *Proc. of the 13th GIF meeting on the Dead Sea Rift as a unique global site* (1997); I. Zak, Evolution of the Dead Sea brines in Niemi In: *The Dead Sea – The Lake and its Setting*, edited by Z. Ben Avraham and Y. Gat (Oxford University Press, 1997), pp. 133–44.
6   A. Flexer, A. Yellin-Dror, J. Kronfeld, E. Rosenthal, Z. Ben Avraham, P. Artsztein and L. Davidson, A Neogene salt body as the primary source of salinity in Lake Kinneret, *Archiv für Hydrobiologie, Beiheft: Ergebnisse der Limnologie* 55 (2000): 69–85.

7   Y. Guttman and E. Rosenthal, Mitzpe Jericho region – research of the salinization mechanism and water potential (Tel Aviv: Tahal, 1991), Report 01/91/46.

8   Y. Guttman, Salinization along the western border of the Jordan Valley and the Dead Sea. In: *Proc. of the 13th GIF meeting on the Dead Sea Rift as a unique global site* (1997).

9   Y. Guttman and H. Zuckerman, The Eastern aquifer in the Judea–Samaria region – calibration of numerical model (Tel Aviv: Tahal, 1995), Report 01/95/66.

10 E. Rosenthal, Ca chloride brines at common outlets of the Bet Shean–Harod multiple aquifer system, Israel, *Journal of Hydrology* 97 (1988): 89–106; E. Rosenthal, G. Weinberger, A. Almogi-Labin and A. Flexer, Late Cretaceous–early Tertiary development of depositional basins in Samaria as a reflection of eastern Mediterranean tectonic evolution, *AAPG Bull* 84 (2000): 997–1014.

11 F. Bender, *Geology of Jordan* (Berlin: Gebrueder Borntaeger, 1974), pp. 182–86.

12 D. Avisar, *Salinization processes and aquifer interconnections in the southeastern Coastal Plan of Israel* (Tel Aviv: Tel Aviv University, 2001), PhD Thesis [in Hebrew, English abstract].

13 Y. Guttman, Salinization along the western border of the Jordan Valley and the Dead Sea. In: *Proc. of the 13th GIF meeting on the Dead Sea Rift as a unique global site* (1997); E. Rosenthal, Ca chloride brines at common outlets of the Bet Shean–Harod multiple aquifer system, Israel, *Journal of Hydrology* 97 (1988): 89–106; A. Vengosh and E. Rosenthal, Saline groundwater in Israel: its bearing on the water crisis in the country, *Journal of Hydrology* 156 (1994): 389–430; A. Flexer, A. Yellin-Dror, J. Kronfeld, E. Rosenthal, Z. Ben Avraham, P. Artsztein and L. Davidson, A Neogene salt body as the primary source of salinity in Lake Kinneret, *Archiv für Hydrobiologie, Beiheft: Ergebnisse der Limnologie* 55 (2000): 69–85.

14 E. Rosenthal, Ca chloride brines at common outlets of the Bet Shean–Harod multiple aquifer system, Israel, *Journal of Hydrology* 97 (1988): 89–106; Y. Yechieli, A. Starinsky and E. Rosenthal, Evolution of brackish groundwater in a typical arid region: Northern Arava Valley, southern Israel, *Applied Geochemistry* 7 (1992): 361–374.

15 Y. Guttman and E. Rosenthal, Mitzpe Jericho region – research of the salinization mechanism and water potential (Tel Aviv: Tahal, 1991), Report 01/91/46.

16 Y. Guttman, Salinization along the western border of the Jordan Valley and the Dead Sea. In: *Proc. of the 13th GIF meeting on the Dead Sea Rift as a unique global site* (1997).

17 Z. B. Begin, A. Ehrlich and Y. Nathan, *Lake Lisan – the Pleistocene precursor of the Dead Sea* (Jerusalem: Geological Survey of Israel, 1974) [in Hebrew, English abstract].

18 Y. Yechieli, A. Starinsky and E. Rosenthal, Evolution of brackish groundwater in a typical arid region: Northern Arava Valley, southern Israel, *Applied Geochemistry* 7 (1992): 361–374.

19 Y. Yechieli, A. Starinsky and E. Rosenthal, Evolution of brackish ground-

water in a typical arid region: Northern Arava Valley, southern Israel, *Applied Geochemistry* 7 (1992): 361–74; E. Rosenthal, R. Nativ and A. Issar, Hydrochemical relationship between rainwater, floods, groundwater and lithology in the Avedat group on the Negev Highlands, Israel. 1984, IAHS. Pub No. 150, pp. 409–418; R. Amit, Shattered gravel in desert reg soils – the effect of salts on the nature and rate of weathering processes (Jerusalem: Hebrew University, 1990), PhD Thesis [in Hebrew, English abstract].

20   J. Drever and C. Smith, Cyclic wetting and drying of the soil zone in an influence on the chemistry of groundwater and terrain, *Am. J. Sci.* 278 (1978): 1448–1454; H. P. Eugster and B. F. Jones, Behavior of major solutes during closed basin brine evolution, *Am. J. Sci.* 279 (1979): 609–631; J. D. Hem, *Study and interpretation of the chemical characteristic of natural water* (USGS Water Supply Paper 2274, 1985), 3rd edition.

21   A. Starinsky, *Relation between Ca – Chloride brines and sedimentary rocks in Israel* (Jerusalem: Hebrew University, 1974), PhD Thesis [in Hebrew, English abstract]; E. Rosenthal, Ca chloride brines at common outlets of the Bet Shean–Harod multiple aquifer system, Israel, *Journal of Hydrology* 97 (1988): 89–106; A. Vengosh and E. Rosenthal, Saline groundwater in Israel: its bearing on the water crisis in the country, *Journal of Hydrology* 156 (1996): 389–430; Y. Guttman, Salinization along the western border of the Jordan Valley and the Dead Sea. In: *Proc. of the 13th GIF meeting on the Dead Sea Rift as a unique global site* (1997); I. Zak, Brines of the Dead Sea Basin: genetic aspects. In: *Proc. of the 13th GIF meeting on the Dead Sea Rift as a unique global site* (1997); I. Zak, "Evolution of the Dead Sea brines in Niemi In: *The Dead Sea – The Lake and its Setting*, edited by Z. Ben Avraham and Y. Gat (Oxford University Press, 1997), pp. 133–144.

22   A. Flexer, A. Yellin-Dror, J. Kronfeld, E. Rosenthal, Z. Ben Avraham, P. Artsztein and L. Davidson, A Neogene salt body as the primary source of salinity in Lake Kinneret, *Archiv für Hydrobiologie, Beiheft: Ergebnisse der Limnologie* 55 (2000): 69–85.

23   A. Flexer, A. Yellin-Dror, J. Kronfeld, E. Rosenthal, Z. Ben Avraham, P. Artsztein and L. Davidson, A Neogene salt body as the primary source of salinity in Lake Kinneret, *Archiv für Hydrobiologie, Beiheft: Ergebnisse der Limnologie* 55 (2000): 69–85.

CHAPTER 9

Lake Kinneret and Water Supply in Israel: Ecological Limits to Operational Supply

K. David Hambright and Tamar Zohary

Lake Kinneret, the Biblical Sea of Galilee (syn. Lake Tiberias), is the only large, natural freshwater lake in much of the Middle East. As such, it has been an important focus of human activity for thousands of years. With the creation of the State of Israel and modern-day development of the country, Lake Kinneret has become the primary reservoir for Israel's National Water Carrier (NWC), an extensive surface and underground freshwater storage and supply network in operation since 1964. In addition to water supply, Lake Kinneret serves important functions in commercial fisheries, recreation, and tourism.[1]

The Long-term Record of Lake Kinneret Water Levels

Lake Kinneret (figure 9.1) is a moderately large (surface area = 169 km²; volume = 4,300 Mm³) freshwater lake located in northern Israel. The distinctive alternations between rainy winters and dry summers character-istic of the local Mediterranean climate produce seasonal fluctuations in water levels which have probably been an integral feature of the Kinneret ecosystem since the present lake formed approximately 20,000 years ago.[2] Minimum water levels were coupled to the amounts of Jordan River inflow and outflow. High flow years led to relatively high minimum water levels; low flow years led to relatively low minimum water levels. Human activity altered lake hydrology since at least 1932, producing four distinct periods of water-level variation:[3] two of natural or semi-natural water level fluctuations (pre-1932, 1948–1972) and two highly regulated eras characterized by increased hydrologic variability (1932–1948, 1973–present).

Based on archeological evidence, Nun[4] concluded that natural water levels in Lake Kinneret fluctuated seasonally over a range of about 1.3m (between –209.5 and –210.8m altitude[5]) between the twelfth and nineteenth centuries (figure 9.2). Nun also suggested that this amplitude of fluctuation, as well as the relatively low maximum water level (–209.5 m), were apparently maintained by the presence of two outlets at the southern end of the lake: a narrow (30–40 m), relatively deep outlet near Degania through which small floods were released, and a wider (150 m) but shallower outlet near Ohalo (see figure 9.1), through which large volumes of water were flushed when the lake level rose due to heavy floods. Over time the Ohalo outlet gradually filled in with sediments and, by the mid-1800s, closed completely. With only the Degania outlet, the maximum water levels in the lake increased by approximately 0.5 m, although, with the exception of the unusual flood years of 1919 and 1929, the overall amplitude of annual fluctuations remained below 1.5 m until the 1930s.

The second distinct hydrologic period began with the construction of a hydroelectric power station south of the lake at Naharayim[6] in 1932. At that time, the Degania outlet was deepened from its natural level of –211 m to –214.5 m altitude and a regulatory dam was constructed to ensure a constant supply of water to the hydroelectric station throughout the year. In order to conserve the lake and its shores from flooding and to prevent over-pumping, maximum and minimum water levels ("red lines") were defined by law under the British Mandate. The "upper red line" was defined as –208.9 m altitude;[7] the "lower red line" as –212 m altitude (figure 9.2). With the operation of the hydroelectric station, the amplitude of water level fluctuations in Lake Kinneret increased from an annual average (± SD) of 1.1 m (± 0.5 m) prior to 1932, to 2.1m (± 0.6 m) for the period 1932–1948, with a maximum fluctuation of 3.2 m in winter 1934/35. The range of water levels recorded for this period was –209 to –212.3m altitude.

An era of semi-natural water-level fluctuations began in 1948, when the hydroelectric station was closed.[8] Nevertheless, the Degania dam continued to operate through the early 1960s, supplying water for agricultural irrigation in the Jordan Valley south of the lake.[9] During this period (1948–1964), annual water-level fluctuations in the lake were gradually reduced to below 1.5 m, while the average annual water level was gradually raised to above –210 m altitude.

Finally, the fourth hydrologic period commenced in 1964 when Lake Kinneret became the primary storage and supply reservoir for Israel's National Water Carrier (NWC).[10] Starting with full operation of the NWC system in 1973, Lake Kinneret water-level fluctuations again increased to greater than 3 m. With Israel's growing needs for water, combined with the capacity to pump water out of the lake through the NWC, pressure was

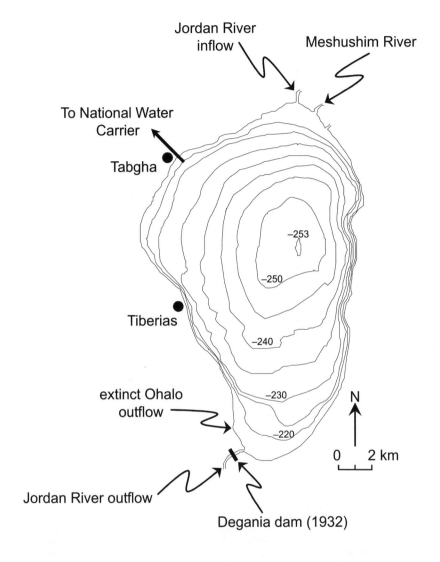

Figure 9.1 A bathymetric map of Lake Kinneret (depth contours are in meters below mean sea level) also indicating the major river inflows, the Jordan River outflow, the site from which water is pumped into the National Water Carrier, the location of the dam at Degania and the old Ohalo outflow.

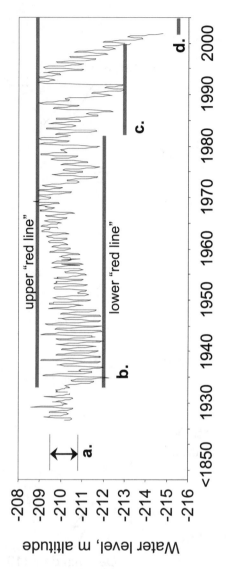

Figure 9.2 A long-term record of Lake Kinneret monthly mean water levels, since the initiation of routine measurements in 1926. **(a)** The 2-sided arrow indicates the range of natural water level fluctuations prior to the 1850s, based on archeological evidence (Nun, 1974). **(b)** The upper and lower "red lines" at –209 m and –212 m altitude respectively, as defined by the British Mandate. **(c)** In 1981 the lower red line was lowered to –213 m altitude, to allow increased pumping during drought years. **(d)** In August 2001 the lower red line was lowered further to –215.5 m, to allow further pumping of water during a period of a regional water shortage crisis.

placed to allow the use of additional water and at the same time increase the storage capacity of the lake. Following much scientific debate over the likely effects of further water level draw-down on the lake's salinity and ecology, in 1981 the Israel Water Commissioner shifted the lower red line by 1 m, to –213 m. Ten years later, in late 1991, and following a two-year drought, Lake Kinneret water level was reduced to its lowest level ever at that time, –213m (figure 9.2).

More due to luck than to proper planning, the winter that followed (1991/92) was the coldest and wettest of the century. The lake filled to its upper red line of –208.9 m (figure 9.2), with excess water flowing down-stream to the southern Jordan River and the Dead Sea. However, a series of average and below average rainfall years from 1994 onwards, combined with increasing water requirements, led to the sequential reduction of the water level of Lake Kinneret from 1994 by about 0.75 m each year, reaching the lowest-ever value of –214.87 m in November 2001 (figure 9.2). Despite continued scientific debate on how the lake will be affected by a further reduction of water levels, a crisis situation dictated the increased pumping, and a further reduction of the lower red line. In August 2001 the Water Commissioner declared the new lower red line to be at –215.5 m, the level below which water can no longer be pumped into the NWC.[11] The last red line reduction to –215.5m has increased the storage capacity[12] of the lake by nearly 400 Mm³ (roughly the current annual pumping rate). However, it is important to note that any water taken from the lake below a level of –213 m is non-renewable in the short-term as it exceeds the natural replen-ishment by rainfall, minus evaporation. Hence, over-exploitation in one year inevitably leads to lower water levels in the following years.

Given the climate of the region, population growth in Israel, and the potential for multi-national use of Lake Kinneret, further reductions in water levels can be anticipated. Although the aesthetic appearance of the lake shoreline and beaches changes drastically with low water levels and large water level fluctuations, the potential environmental impact of these changes on the lake ecosystem are only recently becoming obvious.[13]

Here we provide a brief overview of some of the key ecological implica-tions of management of Lake Kinneret at low water levels. We present evidence suggesting that artificially low water levels may lead to a chain of physical, chemical, and biological changes within the lake, producing symptoms indicative of eutrophication[14] and suggesting an inverse relationship exists between the quantity of water supplied from Lake Kinneret and the quality of that water. We conclude that further attempts to supply more water from the lake will lead to accelerated eutrophication, along with its associated low water quality and high treatment costs. Based on the collective limnological[15] experience with eutrophication accumu-lated worldwide, even a tendency towards accelerated eutrophication of Lake Kinneret should be avoided.

Limnological Consequences of Water Level Reduction

Lake Kinneret is thermally stratified (divided) primarily into two water masses that do not mix except for a brief period in winter (commonly known as "turnover"). An upper, relatively warm water mass (epilimnion) overlies a relatively cooler water mass (hypolimnion) (figure 9.3a). By definition, relative volumes of the epilimnion and hypolimnion are determined by position of the thermocline (an area of sharp temperature gradient between the two water masses), which in turn is determined principally by the amount of heat (as solar incidence) and wind energy applied to the lake's surface.

The epilimnion is primarily the zone of production, in which phytoplankton (algae) convert solar energy, carbon dioxide and nutrients, like phosphorus and nitrogen dissolved in the water, into living biomass, while releasing oxygen to the water in the process. Algae in lakes are analogous to terrestrial plants on land – they are the base of the food web and are thus grazed by herbivores called zooplankton, which, in turn, are consumed by fish. The hypolimnion is the zone of decomposition. Because light from the sun does not penetrate very deep in Lake Kinneret, the hypolimnion is dark and therefore devoid of living algae. Only bacteria can survive in the hypolimnion. The bacteria thrive on the constant "rain" of organic matter (mostly dead and dying algae) being produced in the epilimnion, while consuming oxygen and releasing carbon dioxide and ammonia. In the absence of oxygen production by algae, oxygen consumption by bacteria depletes the hypolimnion of oxygen (see figure 9.3b). By early summer each year all oxygen is consumed, creating conditions conducive to the development of anaerobic bacteria that can obtain oxygen from chemicals commonly dissolved in lake water like nitrates and sulfates. These anaerobic bacterial processes lead to the production of chemically-reduced (lacking oxygen) compounds like nitrogen gas and hydrogen sulfide. Nitrogen gas escapes to the atmosphere, while hydrogen sulfide, along with carbon dioxide and ammonia, accumulates in dissolved form in the hypolimnion.

With no major changes in climate, and hence, the depth of the thermocline (summer average = 15 m), reduced lake water levels will have relatively little effect on the size of the epilimnion, while producing a much smaller hypolimnion (figure 9.3c). For example, a 10 m reduction in water level, from –209 m to –219 m, would yield a lake 37% smaller in volume (from 4,302 to 2,709 million cubic meters). The hypolimnion volume would be reduced by 64%, from 1,989 to 713 million cubic meters, while the volume of the epilimnion would decline only slightly from 2,313 to 1,996 million cubic meters (a 14% reduction). Indeed, during the past three decades there is no significant correlation between water level and ther-

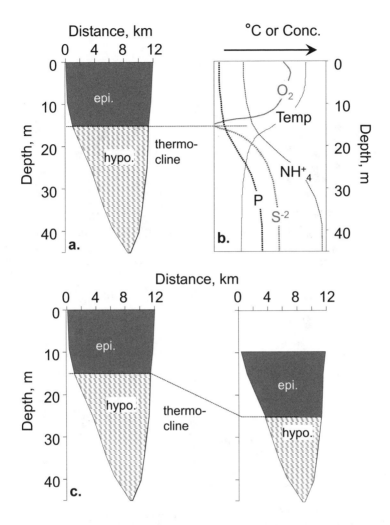

Figure 9.3 **(a)** Schematic diagram showing the thermally stratified Lake Kinneret divided into two water masses: a warm, oxygenated epilimnion overlies a colder and anoxic hypolimnion, with a region of a steep vertical temperature gradient, or thermocline, dividing between the two masses. **(b)** Schematic diagram showing the characteristic changes with depth of temperature, and concentrations of dissolved oxygen (O_2), hydrogen sulfide (S^{-2}), ammonia (NH^+_4), and phosphorus (P) in the thermally stratified water column of Lake Kinneret. **(c)** Schematic diagram showing a 10 m reduction in the water level of Lake Kinneret, from −209 m to −219 m, yielding a lake 37% smaller in volume (from 4,302 to 2,709 million cubic meters). Assuming that the thermocline depth remains at 15 m, the hypolimnion volume would be reduced by 64% (from 1,989 to 713 million cubic meters), while the volume of the epilimnion would decline by only 14% (from 2,313 to 1,996 million cubic meters).

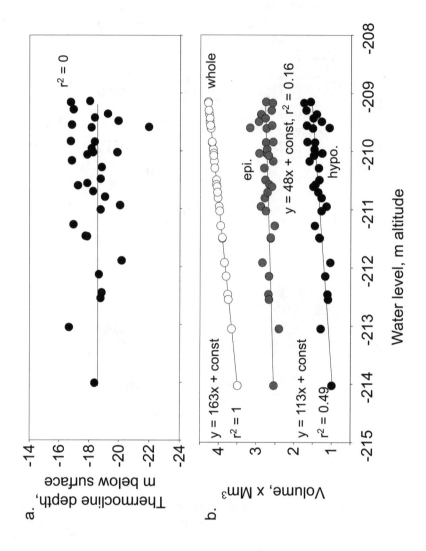

Figure 9.4 Correlations between summer (July to September) mean Lake Kinneret water levels and: **(a)** the mean summer depth of the seasonal thermocline, demonstrating that thermocline depth is independent of water level. **(b)** The mean summer volume of the full lake, the epilimnion and the hypolimnion. This figure demonstrates that hypolimnion volume declines with declining water level at a faster rate than epilimnion volume. The 32 data points on each graph are one for each year, for the period 1969 through 2000.

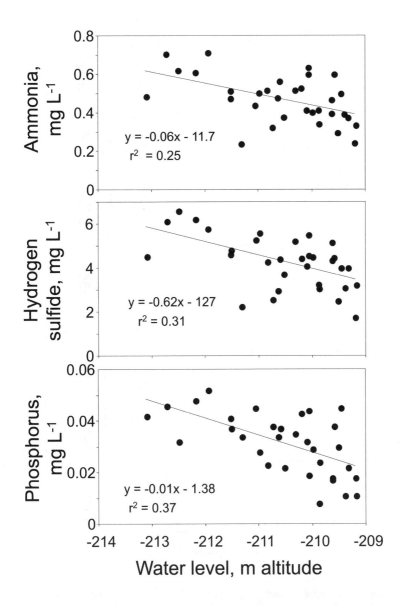

Figure 9.5 Correlations between the summer (July to September) mean Lake Kinneret water level and concentrations of important chemical compounds in the hypolimnion. The average concentrations were calculated from all data points for samples taken from hypolimnetic water in summer. The 32 data points on each graph are one for each year, for the period 1969 through 2000.

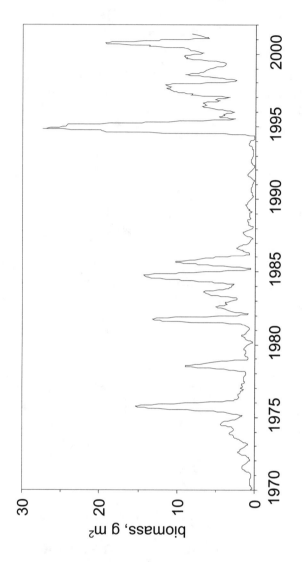

Figure 9.6 Long-term record of bluegreen algal (*Cyanobacteria*) abundance (monthly mean biomass, g m⁻²) in Lake Kinneret during 1970 to 2001. Note the increase in algal biomass since 1994, coincident with the continuous water-level drawdown.

mocline depth (figure 9.4a), and hence epilimnion volume has not changed with declining water levels. However, there is a negative correlation between lake water levels and hypolimnion volumes (figure 9.4b).

Assuming the amount of organic matter produced in the epilimnion is proportional to the size of the epilimnion,[16] a similar amount of organic matter in a low water level lake compared with a high water level lake will sink into, and decompose within, a smaller volume hypolimnion. As a result, higher concentrations of the various end products of decomposition will buildup. Due to a series of chemical reactions commonly associated with this build-up, phosphorus concentrations will also increase in the hypolimnion. Thus, and as confirmed in the 32-year data record for Lake Kinneret, a disproportionately large decline in hypolimnion volume relative to epilimnion volume yields higher hypolimnetic concentrations of ammonia, hydrogen sulfide, and phosphorus (figure 9.5).

As a direct consequence of increased hypolimnetic concentrations of ammonia and phosphorus, whole-lake concentrations of these nutrients will be higher at turnover in winter following a summer of low water levels. These important plant nutrients can stimulate blooms of algae during winter and spring. Although, Berman and co-authors[17] suggest that over the previous 20–25 years there was no major temporal changes in the biomass of Pyrrophyta (the dominant group of phytoplankton in Lake Kinneret during spring), Hambright and co-authors[18] have shown that annual mean non-Pyrrophyta phytoplankton (the dominant groups during the remainder of the year) biomass has increased during this time and is positively correlated with total phosphorus concentrations in the hypolimnion during the previous year. In addition, following analysis of long-term phytoplankton data from Lake Kinneret, Berman and co-authors[19] concluded that seasons with low lake water levels tend to be followed by seasons having higher levels of algal biomass and primary production (the rate at which algal biomass is produced). Indeed, since 1994, there have been dramatic shifts in algal species composition, the most notable and worrisome being an increase in abundance of bluegreen algae[20] (figure 9.6), including toxin-producing species.[21]

Concluding Remarks

Evidence suggests that artificially low water levels may lead to a chain of physical, chemical, and biological changes within the lake producing symptoms indicative of eutrophication. Since 1994, the lake has exhibited uncharacteristic developments in the phytoplankton assemblage, including the first-ever bloom of potentially toxic, N_2-fixing cyanobacteria, thus jeopardizing Lake Kinneret's traditional "good" water quality. Our analyses suggest an inverse relationship exists between the quantity of water supplied from Lake Kinneret and the quality of that water. Further attempts to supply more water from the lake will likely lead to accelerated eutrophication, along with its associated low water quality and high treatment costs. Based on the collective limnological experience with eutrophication accumulated worldwide, even a tendency towards accelerated eutrophication of Lake Kinneret should be a cause for concern.

Water allocation and supply in the Middle East, especially in the Jordan River Valley, is clearly an important topic in the development of regional policies and thus for the Middle East Peace Process. However, the role of Lake Kinneret in solving regional water shortages is limited. Future solutions to the problems associated with water supply must include the development of new water sources, especially desalination of sea water, but should also include improved efficiencies in use of current supplies. It is our obligation to future generations to ensure that Lake Kinneret remains a viable natural ecosystem rather than the over-exploited resource it is at present.

Acknowledgments

We wish to thank all the research scientists at the Yigal Allon Kinneret Limnological Laboratory (Israel Oceanographic and Limnological Research), especially A. Nishri and U. Pollingher for access to the long-term Lake Kinneret Database and M. Schlichter (Database manager) for her assistance.

Notes

1   T. Berman, K. D. Hambright, J. R. Gat, S. Gafney, A. Sukenik and M. M. Tilzer (eds.), *Limnology and Lake Management 2000+. Proceedings of the Kinneret Symposium, September 1998, Ginnosar Israel*, vol. 55 (2000).

2   C. Serruya (ed.), *Lake Kinneret*, vol. 32 (The Hague: Dr. Junk Publishers, 1978).

3   K. D. Hambright and T. Zohary, The history of water level fluctuations in Lake Kinneret, *Ecology and Environment* 2 (1995): 97–100 (in Hebrew).

4   M. Nun, Water levels in Lake Kinneret in the historical period, *Land and Nature* 15 (1974): 212–18 (Hebrew).

5   Note that Lake Kinneret is below mean sea level. Thus natural water levels fluctuated between 209.5 and 210.8 meters below mean sea level.

6   Naharayim is located approximately 7 km south of Lake Kinneret just below the confluence of the Jordan and Yarmouk Rivers. The hydroelectric station was constructed between 1927 and 1932, with inaugural operation of two turbines in 1932. A third turbine was added in 1933. The plant operated primarily with Yarmouk River waters during the winter flood period and primarily with Jordan River waters (via Lake Kinneret storage) during the remainder of the year.

7   Contrary to the opinion stated by Munther Haddadin (chapter 4, this volume), the maximum water level was not increased by 4 m by construction of the Degania dam. Such an increase would have inundated much of the city of Tiberias, as well as many other villages around the lake.

8   During Israel's War of Independence in 1948, the Naharayim Power Station became Jordanian territory and ceased operation.

9   C. Serruya, ed., *Lake Kinneret*, vol. 32 (The Hague: Dr. Junk Publishers, 1978).

10 The idea of Israel's National Water Carrier was first developed in 1944 by W. Lowdermilk (Palestine: Land of Promise, 1944), in which he proposed to use the Jordan River and its tributaries for both electrical energy production and water supply. J. B. Hays (Tennessee Valley Authority on the Jordan, 1948) developed the idea further in technical detail, but Israel's War of Independence and further hostilities between the new state and its neighbors all but scrapped the plan as a potential regional contribution to power and water supply. In 1956, Israel unilaterally implemented portions of these plans relating to its territory, by constructing the National Water Carrier, that today pumps water from the northwest shore of the lake up into the Galilee hills where it flows by gravity as far as the Negev desert. Maximum pumping capacity is approximately 1 Mm^3 daily. Further details can be found in Serruya (1978).

11 The winter rains of 2001–2002 were sufficient for increasing the lake's level only by approximately 1.5 m. Thus, it was expected that the water level would reach –215.5m altitude during the summer of 2002.

12 The storage capacity of the lake is the volume of water contained between the upper and lower red lines. The normal storage capacity between –208.9 and –212m altitude is 504 Mm^3. For lower red lines of –213 m, –214m and –215.5m altitude, the storage capacity is 666, 828, and 1069 Mm^3, respectively. Note that due to high evaporation rates (roughly 300 Mm^3 annually), all of the storage capacity is not available for pumping.

13 A. Parparov and K. D. Hambright, A proposed framework for the management of water quality in arid-region lakes, *Internationale Revue der Gesamten Hydrobiologie* 81 (1996): 435–54; K. D. Hambright, A. R. Parparov and T. Berman, Indices of water quality for sustainable management and conservation of an arid region lake, Lake Kinneret (Sea of Galilee), Israel, *Aquatic Conservation: Marine and Freshwater Ecosystems* 10 (2000): 393–406; T. Berman, Lake Kinneret and its catchment: international pressures and environmental impacts, *Water Policy* 1 (1998): 193–207; K. D. Hambright, M. Gophen and S. Serruya, Influence of long-term climatic changes on the stratification of a subtropical, warm monomictic lake, *Limnology and Oceanography* 39 (1994): 1233–1242.

14 Eutrophication is a process in which excessive nutrient loading, especially of phosphorus and nitrogen, leads to phytoplankton blooms and deterioration of water quality.

15 Limnology is defined as that field of science that centers on the study of inland water bodies, including fresh and saline lakes, reservoirs, streams, and ponds and extends across the disciplines of physics, chemistry, geology and biology.

16 K. D. Hambright, T. Zohary and W. Eckert, Potential influence of low water levels on Lake Kinneret: re-appraisal and modification of an early hypothesis, *Limnologica* 27 (1997): 149–55.

17 T. Berman, Y. Z. Yacobi and U. Pollinger, Lake Kinneret phytoplankton: stability and variability during twenty years (1970–1989), *Aquatic Sciences* 54 (1992): 104–27; T. Berman, L. Stone, Y. Z. Yacobi, B. Kaplan, M. Schlichter, A. Nishri and U. Pollingher, Primary production and phytoplankton in Lake

Kinneret: a long-term record (1972–1993), *Limnology and Oceanography* 40 (1995): 1064–1076.

18   K. D. Hambright, M. Gophen and S. Serruya, Influence of long-term climatic changes on the stratification of a subtropical, warm monomictic lake, *Limnology and Oceanography* 39 (1994): 1233–1242.

19   T. Berman, A. Nishri, A. Parparov, B. Kaplan, S. Chava, M. Schlichter and U. Pollingher, Relationships between water quality parameters and water levels in Lake Kinneret, *Verhandlungen Internationale Vereinigung für Theoretische und Angewandte Limnologie* 26 (1996): 671–74.

20   Bluegreen algae (also called cyanobacteria) often form massive blooms (outgrowths) in lakes that receive high phosphorus loads from the surrounding drainage basin. Such blooms can produced unsightly, bad-smelling surface scums of dead and dying algal cells impeding recreational use and toxins and other noxious chemicals linked to poisoning, carcinoma and other diseases, thus restricting water consumption by livestock, pets and humans.

21   T. Berman, U. Pollingher and T. Zohary, A short history of stability and change in phytoplankton populations in Lake Kinneret, *Israel Journal of Plant Science* 46 (1998): 73–80; U. Pollingher, O. Hadas, Y. Z. Yacobi, T. Zohary and T. Berman, *Aphanizomenon ovalisporum* (Forti) in Lake Kinneret, Israel, *Journal of Plankton Research* 20 (1998): 1321–1339; T. Zohary, Changes to the phytoplankton assemblage of Lake Kinneret after decades of a predictable, repetitive pattern, *Freshwater Biology* 49 (2004):1355–1371.

Part III

Water Economics

CHAPTER 10

The Water Economy of Israel

YOAV KISLEV

The discussion in this essay is influenced by developments that have brought the water sector to its current condition and by the public debate on water in Israel that is overshadowed nowadays by the acute crisis the country is experiencing. Water users are facing recurring shortages and supply to agriculture was cut. One may easily conclude that the sector is in shambles. My review of the issues has led to a different perspective. True, there was neglect, a lot should be repaired, and attention must be paid to changing circumstances; nevertheless the fundamental structure of the sector is sound, a basis for reforms exists, and the water economy can be expected to fulfill its functions, now and in the future.

General Features

Israel is a small and narrow country half of whose area is desert. Precipitation, which occurs mainly in the winter, averages more than 700 mm per year in the north and less than 35 mm in the southern tip of the country. The core functions of the water sector have been to store water from winter to summer and from rainy to dry years, and to carry water from the north to the center and the south. With expanding population and growing urbanization, sewage treatment and recycled water are taking center stage.

Fresh water is stored in Lake Kinneret (Lake Tiberias, figure 10.1) and in several groundwater reservoirs, the two largest being the Mountain Aquifer and the Coastal Aquifer. The Mountain Aquifer is located mostly under the West Bank from a point south of Nazareth to Beer Sheva. The Coastal Aquifer stretches along the Mediterranean Sea from a point south of Haifa to the southern tip of the Gaza Strip. The National Water Carrier

Figure 10.1 A Map of Israel and the National Water Carrier network.
Source: Nurit Kliot, *Water Resources and Conflicts in the Middle East* (London: Routledge, 1994). Political borders are marked in the map by broken dotted lines. The major water arteries are marked by solid lines, the minor arteries by broken lines.

is a system of conduits running west and south from Lake Kinneret and connecting most of the sources and users of water in the country. Two-thirds of the water in Israel is supplied by the largest utility, Mekorot Water Co., Ltd. (hereafter Mekorot), and the company also operates the National Water Carrier. The other suppliers are private well owners, municipalities, and regional cooperatives. Municipalities are required to collect and treat their sewage and several cities have cooperative projects with agricultural interests in their vicinity.

As natural resources, the water reservoirs are common pools. Under open access, individuals will behave as "free riders": they will pump water so long as it is beneficial for their own use disregarding the detrimental effect that their pumping has on other users of the reservoirs (for example, by lowering water levels or drawing in salty ocean water). The common resource will thus be depleted.[1] In addition, suppliers, particularly Mekorot, are monopolies. These features call for government intervention. Consequently, by law, all water sources in the country are publicly owned; there is no private ownership of water. The Water Commissioner is responsible for the utilization and the sustainability of the resources. The law requires measurement of all uses of water. This means that wells and pumps are monitored and consumers – households, manufacturers, farmers, and others – pay according to the quantity they use.

The "safe yield" water supply from natural sources is estimated as 1,550 Mm^3/yr [million m^3 per year].[2] Added to this quantity are 270 Mm^3 of recycled water. This source has been growing in importance in recent years. For example, most of the irrigation water in the western Negev[3] is provided by recycled water from the metropolitan area of Tel Aviv. The forecast is that by the year 2020, Israel will utilize 830 Mm^3 of recycled water per year. The first comparatively large (100 Mm^3 per year) desalination plant on the coast of the Mediterranean Sea south of Tel Aviv commenced production in 2004 and preparations are under way for the installation of additional capacity. Consumption in households and industry was 740 Mm^3 in 2001 and is increasing with population growth. By a government decision, agriculture will be provided with 1,160 Mm^3 per year from all sources with allocation of recycled water gradually expanding until it covers 50% of the supply.

Resources are limited and their development is expensive; population growth has surpassed water supply over the last half-century, and the amount available per person has declined. However, as figure 10.2 demonstrates, per-capita consumption in households and industry has remained essentially constant, while per-person water available for agriculture is today less than half the volume of the 1960s. Despite the reduction, agricultural production per capita is today more than 150% of the quantity produced 40 years ago. By these numbers, water productivity in agriculture has increased three-fold over the period.

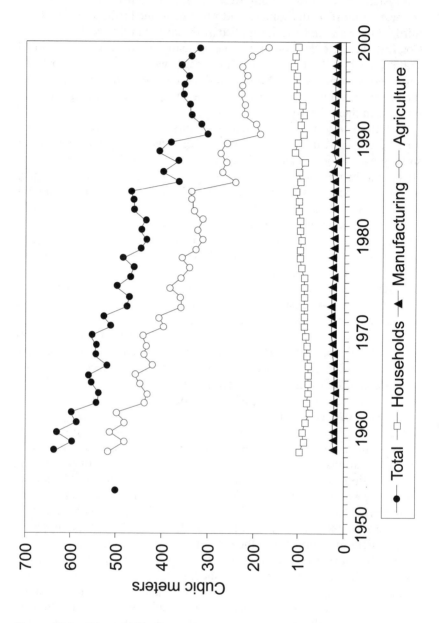

Figure 10.2 Water Utilization (m³ per year) in Israel (per capita) (water use divided by population in the country).
Source: State of Israel, Central Bureau of Statistics, *Statistical Abstract of Israel*, various years.

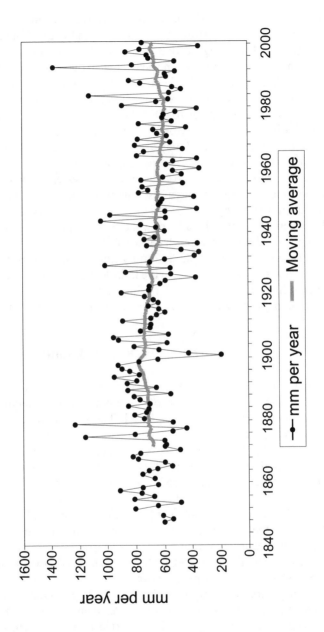

Figure 10.3 Rain in Nablus, mm per year and 25 year, moving average.
Source: Jacob Lomes and Nona Rinborg, Reconstruction of Seasonal Rain in Nablus, 1870–1990, *Water and Irrigation*, 313: 26–29, 1992 (in Hebrew) and data received from The Meteorological Service of Israel. By evidence from other locations, the extremely low value for 1900 is an error.

Water potential and safe yields are generally stated in terms of average use. Precipitation varies from year to year as can be seen from the 150-year record for Nablus[4] in figure 10.3. But not only does annual rain change, long-term averages also vary. The heavy line in the diagram depicts 25-year means; the average for the last quarter of the nineteenth century was more than 20% higher than the lowest value that the heavy line hit – the average for the 25-year period prior to 1979. With varying rain, replenishment of the reservoirs is not stable. Moreover, the changes in the moving average indicate that water potential may also change from one period to another.

There is a tradeoff between average supply and its reliability. With a policy of regular extraction of large quantities of water, reservoirs are often low and, since replenishment varies, reliable supply cannot be maintained. This truism was brought home twice in the last 15 years, once in 1990–91 and again in 1999–2002. Israel is now facing an acute water crisis into which the country slid when several dry years followed a period of over-utilization. The crisis caused a public outcry and even panic. And indeed, in November 2001, the water level in the Lake Kinneret reached its lowest level in known history[5] and the aquifers show clear signs of over-drafting; but the overall picture is not as bad as the public may have been led to believe.

Comprehensive Water Balance

Drawing on Tony Allan,[6] who coined the term "virtual" for water imbedded in traded food, table 10.1 presents a rough calculation of the comprehensive water balance of Israel. The information in the table is based on accepted "norms," not actual data. The norm for average water needs for food production, including rain, is 1000 m^3 per person per year (this is a global average, not specific to Israel). Hence, for a population of 6.5 million, water needed for food production is 6,500 Mm^3 a year. Adding normative values for households and industry, total needs are 7,300 Mm^3 per year. Rain falling on 400,000 hectares of cultivated land contributes 1,600 Mm^3 to the balance, 2,000 Mm^3 come from natural and other sources, and my rough estimate is that 500 Mm^3 of virtual water are withdrawn in exports. The available supply is 3,100 Mm^3. The gap between the needs and available water is 4,200 Mm^3 per year.

The gap is closed by imported food, grain in particular. Israel imports 3.8 million tons of grain a year. On average, again globally, it takes 1 m^3 of water to produce 1 kg of grain; the imported grain is therefore equivalent to 3,800 Mm^3 per year. For the remainder, and to close the balance, I added other imports – sugar, beef, cereals, dried fruits, and others – although I do not know the water content of these products.

The water balance of table 10.1 is just a first approximation and my estimate of local availability was built on optimistic assumptions, but the general picture is clear: water from local sources covers less than 50% of the needs of the country. Moreover, local sources cover only 35% of the water used in food production for the domestic markets. Local water is not crucial for food supply to the population of Israel. This does not mean that we can do without water; it means that a reduction, even a sharp reduction, of allocation to agriculture will not risk the food situation of the country. The crisis is painful but it need not cause panic.

Table 10.1 The comprehensive water balance in Israel (in Mm^3 per year) (see text for details)

Needs	Mm^3 per yr
Food (1000 m³ per person per year)	6,500
Households (100)	650
Industry	150
Total Needs	7,300

Availability	
In the soil (400,000 h × 400 mm)	1,600
Production (including recycled)	2,000
Less: Export	−500
Total Available	4,200

Gap (needs minus availability)	3,100

Import	
Grain (food, feed and oil, 3.8 million t/yr)	3,800
Other	400
Total Import	4,200

Table 10.1 reveals the dependency of Israel on food trade (the country also depends on trade in industrial goods and services). It was pointed out by Allan that this dependency is a common attribute of all the countries in the Middle East; without imports we could not have fed our populations. Starving people would have then fought ceaselessly over every drop of water. Here is globalization contributing to peace.

Hydropolitics

Government intervention in the management of water is a necessary consequence of the common pool features of the resources and the monopoly position of the suppliers. Once the government is involved, in any issue, interest groups arise in an attempt to change public policies in their favor.[7] Lobbies are particularly prevalent in democracies but political pressure can be found everywhere. The two strongest groups in the water sector, the farmers and the workers of Mekorot, have different interests and attempt to affect different aspects of policy. (The "greens" form the third group. They are growing in strength but their effect is still marginal.)

The main interest of the farmers is to get large supplies of water at the lowest possible price. Water is an important input in agriculture and many farmers enthusiastically support their representatives in the political arena. The agricultural lobby is therefore well-organized and acts vigorously in advancing its case. It is said in Israel that a politician stands firm when pushed on all sides.[8] But, as water is not an economically important item in the budget of households or in the cost of manufacturing, the farmers do not face strong opposition and they have succeeded in tilting the policy in their favor. It must be added, however, that the farm lobby, although still alive and supported in high places, has lost some of its power in the last decade or two. The loss can be attributed to the decline of the share of agricultural output and employment in the national economy, as industry and services expanded, and perhaps also to the growing intuitive comprehension, by the public, of the realities presented in table 10.1.

Comparatively low prices and large quantities of water allocated to agriculture result in two major consequences: (a) the reservoirs are depleted and supply is put at risk; and (b) economic waste is created in the sense that water is used in products that cannot cover the real cost of this factor of production. This is particularly true for "virtual" water in some of the exports; citrus is an example.

Interest groups, farmers among them, do not adopt the point of view of the economy at large – they act as "free riders". The farm lobby advocates expensive expansion of supply to overcome shortage and it is quick to find patriotic justifications for subsidies to agriculture. The ability of a lobby to affect policy, the power of the interest group, depends on the environment in which it operates. Consider a region drawing water from a river and exhausting the flow, both in regular and in dry years. There is no more water than what the river carries and, as much as it may try, a political lobby cannot pressure the authorities to increase allocation. The situation is different when supply is, as in Israel, from reservoirs. The total amount of water stored in the coastal aquifer, to take one example, is estimated to be 18,000 Mm^3; safe yield is 250–300 Mm^3 per year, less than 2% of the storage. Over-drafting accelerates the accumulation of salts in the aquifer, but there is no physical constraint to tie the hand of the Water Commissioner when he, willingly or reluctantly, yields to pressure for more water.[9]

Originally, the Water Commission operated from the Ministry of

Agriculture and was deemed to be under the influence of agricultural interests. The Water Commission was moved, several years ago, to the Ministry of National Infrastructures. The change was not motivated by efficiency considerations but it was welcomed by many observers of the water sector. Their happiness was premature. The Minister of National Infrastructures at the time, Ariel Sharon, was a farmer; he even owned the largest private farm in the country, and he appointed another farmer, Meir Ben-Meir, as the Water Commissioner. In my judgment, it was the pro-farm policy of this commissioner – a myopic pro-farm policy – that paved the way for the current crisis.

The workers of Mekorot form the other interest group in the water sector. Their power stems from their organization – 2,200 men and women under a strong union leadership – but especially from their control of the supply: their hand is on the tap. The first interest of the workers is income; their salaries are among the highest in the country. Inflated salaries increase the cost of water, but perhaps more costly is the support the workers give to the monopolistic power of the company. As a large monopoly, Mekorot may build expensive projects and secure employment for its workforce, safely assuming that all costs will eventually be covered. The government has been trying, for more than 10 years, to reform the company, to separate its operations into several relatively small units, and make it into a public utility, independent and responsible for its finance. The workers opposed the reform and, although it did have some important effects on the economic functioning of the company, the reform was only partly successful. An indicator of the remaining strength of Mekorot is that the present Water Commissioner (in office since mid-2000) was the chief engineer of Mekorot before assuming his position as a public servant. (Meanwhile, the former Minister of National Infrastructures became the Prime Minister and has been using his top position to block austerity moves suggested by the new Water Commissioner.)

Allocation

There are two major allocation problems in the water sector: (a) allocation of extraction – where, when, and how much to pump; and (b) allocation of water for utilization and consumption. The two problems are distinct, although the law obscures the distinction.

The criterion for extraction of water is sustainability of the resource. The role of the Water Commissioner is to guard the long-term stability of the quantity and quality of water. Fulfilling this role may require decisions on each source and well separately, depending on local hydro-geological circumstances. Accordingly, the law specifies that water may be extracted only under a license from the Water Commissioner.

The criterion for the allocation of water for consumption and utilization is efficiency, that is, the maximization of economic welfare from the use of water. Two management instruments are in use: prices and quotas. Households and most manufacturers can purchase from Mekorot all their demand at the established prices. Water in agriculture is allocated by quota and, in addition, farmers who purchase water[10] pay Mekorot or the regional suppliers. In principle, quotas are reallocated every year; in practice, they have not changed much in the last several decades (marginal changes were made and will be discussed below).

Prices reflect cost. Two major factors have affected cost of water in Israel: one has been the shift from relatively inexpensive to higher cost systems and the other has been the rise in the world price of energy since 1974. I start with a sketch of the historical development of cost and demand in figure 10.4. The diagram is drawn in today's prices; that is, it does not reflect past changes in the price of energy. The stepwise increasing graph traces my assessment of the cost of water: cost of local supply is 12 US cents per m³, average cost of water supplied via the National Water Carrier is 35 cents per m³, and cost of desalination is estimated to be 60 cents per m³.[11] The graph also traces historical changes of costs in the three epochs of the development of the water economy of Israel. In the 1950s and early 1960s Israel was in the epoch of local supply; the National Water Carrier opened in 1964 a new epoch of expansion, at a higher cost, from 600 Mm³ to 1,550 Mm³ a year. The period beginning from 2004 is the epoch of desalination.

The lines marked 1960, 1970, 2003, and 2010 represent demand in these years. In 1960, a few years after the establishment of the State of Israel, irrigation was not widely practiced and the demand for water was modest. When the cost to users was 12 cents the quantity used was less than the potential supply of the time. But it has expanded. In the early years the expansion was generally due to the introduction of irrigation into areas of dry farming, and more recently, when the expansion of agriculture has been slower, most of the increased demand is in urban areas, due to growth of population. As plotted, the demand of 2003 put Israel in the transition period, between the epoch of the National Water Carrier and that of desalination. We shall return below to some of the implications of the conceptual framework of figure 10.4.

The government sets the price that the controlled monopoly Mekorot may charge; essentially it is the same price for all users in agriculture. Regional suppliers – most of them cooperatives – charge to cover cost. In the past, the Water Commission operated an Equalization Fund: well owners and other low-cost suppliers (who had access to local sources of water) contributed to the fund, high-cost operators were compensated. Since Mekorot was the major provider of water to remote and hilly areas, the company received the lion's share of the accumulated funds.

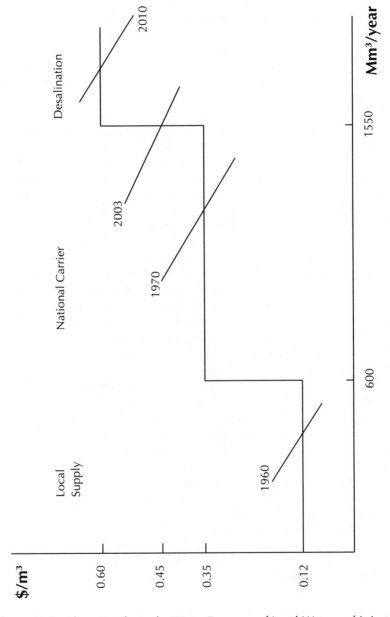

Figure 10.4 Three Epochs in the Water Economy of Israel Water and Irrigated Products (1952 = 100).
Source: State of Israel, Central Bureau of Statistics, *Statistical Abstract of Israel*, various years.

The index of real prices of water (deflated by the consumer price index) is depicted in figure 10.5. The index aggregates prices paid to Mekorot and to other suppliers. Since 1952 the price of water has increased two and a half times. But, as the graph clearly indicates, water prices did not increase markedly in the 1960s when water from the more expensive National Water Carrier was first delivered to the south. Price increases were made politically feasible by the dramatic rise in energy costs, in the early 1970s. This observation helps to explain the different experiences of Israel and the western United States.

It was recently reported[12] that farmers in Central Utah paid 0.7 cents per m^3 while the value of the marginal productivity of water was 2.5 cents per m^3, and the cost of providing the water was 25 cents per m^3. Farmers in California paid 1.2 cents per m^3. The American prices are much lower and the gap between cost and user pay is much wider than the corresponding values common in Israel. The explanation may lie in the nature of the supply. Most of the water supplied by the large projects in the Western United States is captured in dams and moved by gravity. The major component of cost is capital; this is sunk cost (i.e., cost that cannot be recovered) and, with the conventional accounting practices, it does not figure in the current public budgets. In contrast, Mekorot elevates water from Lake Kinneret, 215 meters below sea level, close to 400 meters to the hills and pushes it southwards. Israel's is therefore an energy intensive water project and the agricultural political lobby could not prevent a Treasury burdened by increasing energy bills from making Mekorot's consumers share in the cost. Private and regional suppliers were affected by the increasing energy prices directly, in their electricity bills.

The other graph in figure 10.5 depicts the index of the real prices farmers received for products of irrigated agriculture (crops, horticultural products, and flowers). Since the mid-1950s, except for the decade of the 1970s, the index has shown a downward trend, reflecting developments in world markets where abundant supply has historically reduced food prices.[13] Farmers in Israel were caught in the scissors action of rising water prices and decreasing product prices. The consequences are clearly seen in figure 10.6: up to the mid-1980s water used by farmers exceeded the quota (aggregate quota of the sector) but, since then, agriculture has used less water than the quota allowed. True, some farmers may have been limited by the quota, but for the sector as a whole, in the last two decades, water was allocated by prices – of water and products – and not by the administrative quota.

As indicated earlier, most of the quotas have been stable since the early 1970s. The dips in the mid-1980s and in 1990 are reflections of temporary cuts in allocation to agriculture when shortages occurred; particularly severe was the second crisis. The crisis of 1999–2002 is still not reflected in the figure. For many farmers, quotas did not change, but not for all. There has been a persistent increase in aggregate quota throughout the period

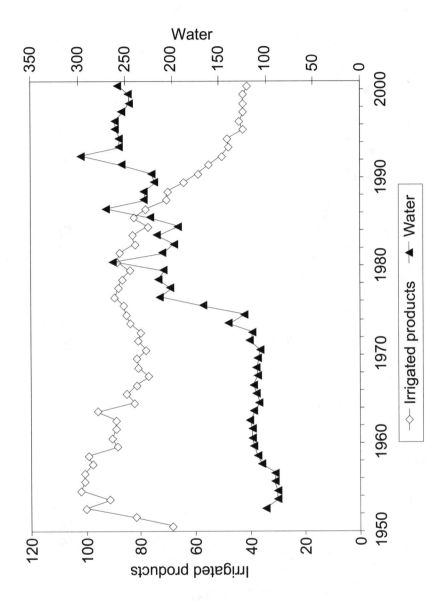

Figure 10.5 Price Indices in Agriculture, Water and Irritated Products (1952). *Source*: State of Israel, Central Bureau of Statistics, *Statistical Abstract of Israel*, various years.

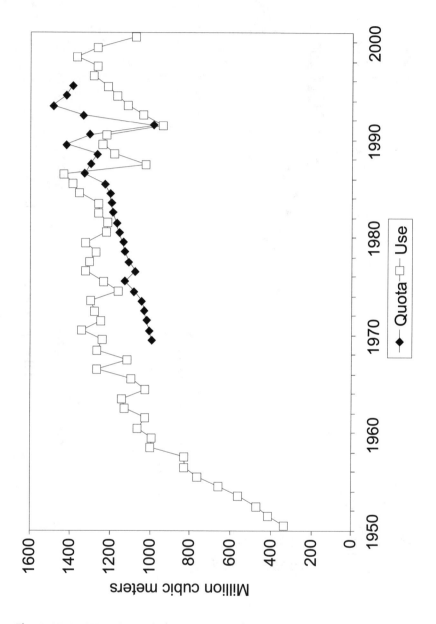

Figure 10.6   Water in Agriculture, quota and use (Mm³ per year).
Source: For uses: State of Israel, Central Bureau of Statistics, *Statistical Abstract of Israel*, various years. For quota: Assaf Perez, *Water Market Regulations in Israel: Quota, Prices, and Political Lobbying* (Rehovot: Hebrew University of Jerusalem, MSc Thesis, 1999).

covered in the diagram. The crisis of 1990–91, like the current one, was created by a period of drought following years of over-drafting and tight supply. Still, allocation to agriculture kept expanding and the expansion policy was myopic: the cost of the crisis may turn out to be higher than the cost of gradual and controlled reduction in supply.

The increased allocation to agriculture encouraged use of water in the sector (we shall soon see how) despite gradual growth in fixed demand for urban water. The resulting over-drafting need not be large (estimates vary); even the smallest annual gaps accumulate, as every household with a bank account knows, whether in theory or from experience. A jackpot may however save, at least for a while, a financially careless family. Israel hit a jackpot in the winter of 1991–92 when rains surpassed all records (figure 10.3), reservoirs were filled to capacity and the books were reopened on blank pages. The policy of over-drafting of the last several years is evidence of a refusal to recognize that the roulette seldom returns a second time to favor lavishly the same player.

Prices and Subsidies

Water prices represented by the index of figure 10.5 were average payments per m³. Mekorot customers in agriculture pay block rate prices that rise with the amount purchased: 20 cents per m³ for any quantity up to 50% of the quota (the first block), 25 cents for the next 30% and 32 cents for additional amounts. The purpose of this price structure is to combine support to agriculture with economic efficiency: support comes through lower prices for part of the amount used and efficiency is allegedly assured due to the fact that the marginal unit is paid at full price. Although the introduction of block rate pricing in 1989 had some economic benefits, the claim that it is an efficient pricing system is flawed in more than one way.[14] We shall not go here into technical details, but will consider one effect of the method: by increasing the quota of a farm, as was occasionally done (figure 10.6), the Water Commissioner extended the range of lower blocks. In this way he reduced the average price that the farmer paid and increased the support the farm received. Block rate pricing thus gave the Water Commissioner powers that the legislators of the original Water Law did not have in mind. Moreover, in expanding the quota, the Water Commissioner encouraged use of water by the fortunate farmers who got the new allocations, even if the aggregate quantity was determined by prices.

Economic support must be covered somehow and, indeed, the government covers regularly 15–20% of the cost of Mekorot.[15] But this is not the whole picture; municipalities pay Mekorot "at the city gate" a wholesale price of 35 cents per m³, while the average price the farmers pay is only 25 cents per m³. In this way, the urban consumer cross-subsidizes water in agriculture. There is, however, another way to look at the price structure and the cross-subsidization it implies. As indicated earlier, all farmers pay the same price and similarly municipalities pay one price. But costs vary; farmers in the far south and in the hills pay less than the cost of supply to their areas, while farmers in the north and center pay more than the cost. Similarly, Tel Aviv, located on the coast, pays more and Jerusalem, at a higher elevation, pays less than its specific cost of water provision. Economic considerations would call for adjustment of prices to cost, but I cannot see the politician who will survive even a mild support of a rise in water rates in Jerusalem, the holy capital, relative to Tel Aviv, the city of business and entertainment.

Economic considerations not withstanding, governments often subsidize business and other activities. Water is a convenient medium of support; it carries the subsidy to the end of the pipeline, where costs are relatively high and farming conditions are harsh. Support of sectors in need is a government prerogative but good public housekeeping demands that costs be calculated properly and subsidies indicated explicitly in the budgets. General "principles" of one price, which implicitly support some users more than others, obscure policies and reduce their effectiveness.

Extraction Levies

Another, new form of price is the *extraction levy*, which was imposed in 2000.[16] Consider figure 10.4: as the demand for the year 2003 is depicted, consumers – farmers and others – are willing to pay more for water than the pecuniary cost of 35 cents per m³. In the diagram, if the price for 2003 is set at 35 cents, the quantity the consumers of all sectors will try to use exceeds the available amount of 1,550 Mm³. The marginal value of water to its users is, in the diagram, 45 cents and this will be the equilibrium price equating supply and demand. The difference between the marginal value of the water and the direct cost of supply is termed scarcity value. The extraction levy is set to reflect the scarcity value. If Mekorot is charged 10 cents as the extraction levy for water delivered through the National Water Carrier, the cost of the company, and the price the consumers will pay, will be the equilibrium price of 45 cents per m³. The scarcity value of local water is even higher and so also will be the extraction levy. This is the idea behind the extraction levy; the tax is calculated to reflect the "scarcity value" of water. It is imposed on well operators, including Mekorot, and it varies by region according to specific regional scarcity values. The current rates are between 10 and 13 cents per m³.

This is not the place to go into a detailed discussion of the pros and cons of the extraction levy, but it is appropriate to emphasize that the levy is an important economic instrument. When the levy is in effect, water users are confronted by the scarcity values. The levy makes users take the social effect of individual pumping into their own private considerations and, with it, they behave less as "free riders" and more as considerate users of a scarce resource. Although pump taxes are used in some places, I do not know of any country outside Israel that has introduced scarcity prices for water.

Needless to say, farmers opposed the extraction levy vehemently. It passed, despite their opposition, with the help of political horse-trading. Initially it was just a change of title; payments to the Equalization Fund were renamed extraction levies, and the farm lobby was "bribed" by a promise that all money accumulated in the defunct fund will go to support recycling projects. Once the levy was established, it could be raised and expanded to regions that were originally not covered by it. Also, evidently, the passage of the levy is an indication of the waning political power of agricultural interests.

The shift from equalization contributions to extraction levies was not just a change of title. In principle, there is an important difference between the two. Equalization charges were assessed on every supplier and well owner individually according to their specific operating costs. The charges encouraged "tax planning" and investments that were not economically justified. Extraction levies should not be affected by the cost of the individual supplier.[17] Payments to the Equalization Fund distorted economic incentives; the extraction levies improve them.

Sewage and Recycled Water

By recent estimates, 60% of the water used in households and industry may be returned as sewage.[18] 80% of the sewage is now collected and treated. Reclaimed water adds 30% to fresh water used in agriculture. This ratio is expected to increase both as the allocation of potable water to farm use is decreased and as sewage collection and diversion of recycled water to agriculture intensify.

As indicated above, the average cost of water to urban areas is 35 cents per m³. The cost of distributing the water to households and businesses, collecting the sewage, treating it, and getting the recycled water to the farms can add between one and two dollars per m³. Traditionally, we have regarded the water economy as consisting of the extraction and conveyance systems and viewed the urban water system as its small appendix. But by now, the urban water economy from the city gate to the consumers, to the treatment plant, and to final disposition is not smaller than the economy of fresh water; and it is growing.

The government supports the sewage sector at two levels. At the first, the government finances investment in sewage and recycling projects in municipalities. This line of support from the state to the local authorities grew markedly in the 1990s when the impossibility of enforcing the law requiring municipalities to collect and treat their sewage was recognized. City managers found it so much easier to let the waste flow into the nearest streambed rather than investing in expensive treatment systems.

At the second level, the government supports investment in the adaptation of irrigation to reclaimed water. The cost of adaptation is not negligible; storage is prepared to keep treated water from winter to summer, and new networks are constructed to assure that recycled sewage is not mixed with drinking water. I do not know of assessments of the value of the subsidies entailed by government support to sewage and recycling activities, but essentially most of the initial capital outlays are covered by public funds. Farmers and their regional cooperatives cannot raise on their own the amounts needed for these projects on the capital market.

Judged by its size, the Tel Aviv treatment plant with the pipeline moving its product – high quality reclaimed water – to the Negev in the south is regarded as a national project and operated by Mekorot. For the other places, the accepted doctrine has been to encourage local solutions. In most cases, regional cooperatives take the recycled water from nearby cities. However, local solutions to the sewage problem raise difficult economic questions; for example, should farmers pay the cities for the treated water? Or, how can the cities be assured that the farmers will not reduce usage one day and leave the municipalities with treated water of which they have no way to dispose? The recognition of these problems was one of the motivations for a fresh examination of the national sewage problem in a new master plan,[19] a second motivation will be explained below. The examination will include a comparison of the doctrine of local solutions against the alternative of a central conduit that will carry treated sewage from the coastal region to the Negev.

Quality Issues

Several sources add salts to the water reservoirs in Israel; we shall consider two examples. In the north, salty springs flow into Lake Kinneret; in the coastal area, winds deposit on the ground drops of water carried from the Mediterranean Sea and the rains drain the salt into the aquifer. In the past, under natural conditions, when water was not extracted from the reservoirs, equilibrium prevailed; on average, a certain amount of salt was added yearly and the same amount was withdrawn. The process was visible in Lake Kinneret: nearby springs of brackish water added salt to the lake and the Jordan River drained it towards the Dead Sea. A similar process operated under the ground in the aquifers. Under natural conditions, on average, the same amount of water added yearly by rain to the aquifer (the replenishment) flowed into the Mediterranean Sea. The flow carried with it, again on average, the same amount of salt as was added annually.

Identical, on average, input and output flows kept the salt content of the reservoirs constant. Thus 100 years ago salt concentration in the Coastal Aquifer was some 60 mg of Cl per liter (chloride is an easily-measured and common proxy for salt content) and the water was regarded as having high quality. The natural salt content of Lake Kinneret was approximately 350 mg Cl per liter. As part of the construction of the National Water Carrier, several of the salty springs were diverted to the Jordan River south of the lake (in the Chloride [Saline] Water Carrier; figure 10.1) and, as a result, the salt concentration was reduced to 230 mg Cl per liter,[20] a level considered tolerable for most crops. Today, when water from Lake Kinneret is pumped to the National Water Carrier, salt from springs that were not captured and diverted is also carried in the pipelines. In this way, the concentration of salt in the lake is kept from increasing.

The situation in the Coastal Aquifer is different. In the epoch of local supply (figure 10.4), wells pumped water from the aquifer for irrigation above it. The flow from the aquifer to the ocean was reduced and most of the irrigation water evaporated. Part of the salts carrying water filtered to the subsoil and the groundwater. Four sources have further added salts and other pollutants to the coastal aquifer in the last several decades. One was irrigation over the aquifer with water from Lake Kinneret; the second was heavy use of fertilizers in agriculture that resulted in some chemicals, particularly nitrogen, leaking into the groundwater; the third source was salts seeping into the aquifer when extraction reduced water-level, thus lowering the pressure that had kept salty water at bay; and the fourth source was urbanization with leaking sewage, oil, and industrial pollutants. The average salt content of the Coastal Aquifer has reached 200 mg Cl per liter and is rising. More than a few wells are not operated anymore because of particularly high local concentration of pollutants. The situation is even worse in the Gaza Strip and in the smaller aquifers north of the Coastal Plain.

Pollution and water quality are the second motivation for the fresh examination of the sewage problem mentioned in the previous section. Treated effluents are significantly saltier than the background water (water used in households and businesses from which the sewage was collected) and, even after treatment, the water may be polluted with other undesired chemical and biological ingredients. The use of reclaimed water, depending on the degree of treatment, is therefore limited to insensitive crops. In addition, irrigation with recycled water above aquifers may pollute the underlying reservoirs rendering their water unsuitable for home use and, in the long run, also for agriculture. The Water Law empowers the Water

Commissioner to act in preventing pollution, but strict regulations have not been enacted as of yet. It seems that the regulator and his advisers are seeking compromises between the needs of sustainable aquifers and policies the farmers can live with.

Municipal Water Services

Municipalities get water from their own wells or they purchase it from Mekorot. They are responsible for delivery to households and businesses and for the collection of sewage and its treatment. Prices paid by consumers, including a sewage charge, are set by the government to cover municipal costs. The price is calculated to cover cost of depreciation and maintenance components but local politicians – not unlike their colleagues at higher levels of government – are short-run maximizers; the municipalities neglect maintenance and use the funds saved to finance visible, aboveground projects. The consequences are obsolete networks, water losses, and local authorities whose interest lies, not in saving, but in increasing water use. The government is attempting to privatize municipal water supply; the model often referred to is Buenos Aires, where a French company took over the city's water services.[21] The change is slow; municipalities are reluctant to lose the goose that lays the golden eggs.

Water Markets

By the wording and the spirit of the law, users cannot sell their water rights. In reality there has always been substantial trade in water, some of it officially sanctioned and some without the knowledge of the Water Commissioner. Moreover, in the cooperative villages (moshavim) the quota is allotted to the villages, and farmers may buy from the village pool or sometimes sell to it. Consequently, at present, some farmers have access to traded water while others are barred from it.

The justification of the official anti-trade attitude was the need to maintain flexible policies that can be modified as circumstances change. It was thought that markets would strengthen the hold of private parties on their water, and the property rights so created will stand in the way of proper management of the resources. This consideration was somewhat theoretical; in practice, even now, when trade is officially forbidden, stable quotas are taken as belonging to the farmers and cannot be modified arbitrarily ("possession is nine points of the law"). Across the board cuts are acceptable in emergencies, such as in time of drought. Justly so, prices are the most efficient instrument of allocation in regular times but they are not the right means in emergency situations. It is impossible to determine the exact price that will cut water use by the needed amount for one or two years, and increasing the price of water in time of shortage will ignite strong political opposition – "not only do we have less water but the price also rises!"

Emergency situations are, however, the appropriate time for trade in water. Indeed, water markets in California were most active in dry years.[22] When quotas are cut administratively, some farmers are left with allocations they do not need and others are under-supplied. Trade may amend this deficiency. Unfortunately, the Ministry of Agriculture and the Water Commissioner joined forces and announced recently that water transfers were strictly forbidden in time of crisis. The inability to enforce the restrictive policy can be expected to change the underlying philosophy.

Rules vs. Discretion

The law gives the Water Commissioner (and the minister above him) a set of powerful instruments to enforce the chosen policy. The law does not, however, specify the policy or the duties of the commissioner. The implication of the omission is that the lawmakers trusted the Water Commissioner to manage by discretion, to use professional judgment in formulating policies and directing the sector. Experience taught, however, that management by discretion failed. Throughout the years, the Water Commissioner allowed over-drafting, the sector was brought at least twice to a severe crisis, and aquifers have been depleted and polluted. This need not be so. Management may be by rule whereby mathematical and economic models[23] can be formulated to calculate alternative rules and to assess their effects, such as whether and by how much to curtail supply to agriculture in a dry year, or how much water to allocate in a rainy year. Once the principle of management by rules is accepted, the government will be presented with the set of alternative rules and their implications and it will adopt a policy of choice. The performance of the Water Commissioner will be judged by his adherence to the rules and by the pre-specified goals achieved.

It may seem that the objective of the shift to management by rule is to clip the wings of the Water Commissioner. In some degree it is, particularly in light of past experience. But this effect should not be exaggerated; a Water Commissioner under strong political pressure is not that independent, and his wings are not stretched very wide. Moreover, water management will always be management under uncertainty. With management by rule the Water Commissioner is relieved of part of the responsibility for risky, often painful, policies.

Concluding Remarks

In the opening of the survey I expressed my opinion that the basic structure of the water economy of Israel was sound. The question is: what changes and reforms are now needed, particularly in view of the current crisis and the policies that led to it?

The crucial action to take immediately is to reduce sharply water supply to agriculture (this will naturally entail heavy compensation of affected farmers). There is no other way; the first desalination plant on the Mediterranean Sea, operating since 2004, adds only 100 Mm^3 yearly – less than 5% – to the country's water supply. Other plants will be even slower to come. Unless there is an exceptionally rainy year, following a policy of mild reductions in water allocation to agriculture will cause severe and irreparable damage to Israel's water resources. All experts agree on the damage; some assert, however, that the Coastal Aquifer is already lost as a high quality reservoir and in the future its water will require purification and desalinization. The aquifer, they conclude, can now be mined to keep agricultural water use at the current level until desalination in large quantities replaces it.

The mining possibility illustrates clearly the necessity that the government (I mean the cabinet, the highest level) becomes involved in the water economy more than it has been in the recent past (the governments of the 1950s were very much involved). The mining of the Costal Aquifer will have large financial and political implications and the decision cannot be left to the administrative level. And this is not the only question of major importance that the water sector is now facing. Several other issues are waiting to be resolved with a long run perspective in mind. This is the motivation for my suggestion above of management by rule.

Israel has the expertise and the legal and administrative basis to execute the needed reforms. But the reforms cannot be taken for granted; governments have short attention spans, and they have to be pushed. The public and the media will push if they are informed. A great deal of information on the water economy is constantly accumulating, but most of it is of a technical nature, not accessible to the public at large. The policy and its underlying considerations must be made open and explained to the public.

Reforms hurt vested interests; politically they are accepted only in severe and painful crises. One hopes that the current crisis is painful enough. Once the sector is reformed, the public discussion of water will be more rational than it has been to date. This will also open the way for a rational discussion of regional water issues, matters of war and peace in the Middle East and the subject of our gathering for this conference.

Acknowledgments

Eli Feinerman and Yakir Plessner read an earlier draft and offered helpful comments. David Hambright edited the chapter carefully. Discussions with Gadi Rosenthal added insight and information. Devoira Auerbach helped solve language problems. The responsibility for the product is mine.

Notes

1   For an excellent discussion of common resources, see G. Hardin, "The Tragedy of the Commons," *Science* 162 (1968): 1243–48.

2   The information is from The State of Israel, The Water Commission, *The Tasks of the Water Economy of Israel in the Long Run* (Tel Aviv: 2000, in Hebrew). The available supply of 1,550 Mm^3 per year is of water to use in Israel, the Golan Heights (north-east and east of the Lake Kinneret), and the Jewish settlements in the West Bank and Gaza Strip. Earlier, higher estimates (a number often cited was 1,830 Mm^3) included water lost in floods, used in the West Bank, and supplied now to Jordan.

3   The Negev is the southern part of the country. In figure 10.1, roughly the area south of a straight line from Hebron to Gaza.

4   Nablus is not in Israel proper; it is located in the West Bank, in the territory controlled now by the Palestinian Authority (figure 10.1). It can, however, be taken as representative for the northern, rainy parts of Israel.

5   Chapter 9, in this volume.

6   Tony Allan, *The Middle East Water Question* (London: I. B. Tauris, 2000).

7   For a theoretical exposition of political lobbying in the water economy, see Pinhas Zusman, "Informational Imperfections in Water Resource Systems and the Political Economy of Water Supply and Pricing in Israel," in: Douglas D. Parker and Yaacov Tsur (eds.), *Decentralization and Coordination of Water Resource Management* (Boston: Kluwer Academic Press, 1997), pp. 133–54.

8   The assertion can be proved mathematically in a model of political lobbying. See Gene M. Grossman and Elhanan Helpman, "Protection for Sale," *American Economic Review* (1994): 84 (4), 833–50.

9   Overdrafting of aquifers is not unique to Israel. "Virtually everywhere, governments and farmers have their heads in the sand on the groundwater problem – but it is not going away. Irrigation cutbacks will occur." Sandra Postel, *Pillar of Sand: Can the Irrigation Miracle Last?* (New York: W. W. Norton & Company, 1999), p. 251.

10   Some farmers own private wells and others, around and above Lake Kinneret, pump directly from the lake or the Jordan River (all under license).

11   The price implicit in the contract signed recently with the company chosen to build the first desalination plant was surprisingly low, 53 cents per m^3. Connecting to the National Water Carrier will increase the cost somewhat.

12   Erin Schiller and Elizabeth Fowler, *Ending California Water Crisis: A Market Solution to the Politics of Water* (San Francisco: Pacific Research Institute, 1999) (downloaded from <www.pacificresearch.org>).

13   Robert W. Fogel, "Catching up with the Economy," *American Economic Review* (1999): 89 (1): 1–21.

14 Ziv Bar-Shira and Israel Finkelshtain, "The Long-Run Inefficiency of Block-Rate Pricing," *Natural Resource Modeling* (2000): 13(3), 471–92.

15 Cost and subsidy information in this passage is from the budget of Mekorot Water Co., Ltd, various years.

16 Takdin, Israel laws and regulations, "Water Regulations (Extraction Levies) 2000" (Compact Disk, 2001, in Hebrew).

17 The difference is not complete; the levy may be lowered for high cost private (non-Mekorot) suppliers. It is not known yet how important this provision will be.

18 Water Commission, "The Tasks," p. 6. This assessment seems however to be somewhat optimistic.

19 Tahal, Consulting Engineers, Ltd, *A Master Plan for Effluences in the Center and the South of Israel* (Tel Aviv: 2001, in Hebrew).

20 Executive Action Team (EXACT), Middle East Water Data Banks Project, *Overview of Middle East Water Resources* (U.S. Government Printing Office (1999), p. 31. As part of the peace treaty with Jordan, Israel promised to desalinate the water in the Saline Water Carrier and to supply part of it to Jordan, but this has not been done yet.

21 Lorena Acazar, Manuel A. Abdala, and Mary M. Shrily, "The Buenos Aires Water Concession," in: <econ.worldbank.org/docs/1065.pdf> (n.d.).

22 Richard J. McCann and Eric Cutter, "California's Evolving Water Markets, A Case Study from 1977 to 2000" (2002): University of California at Davis, mimeograph.

23 One possibility is dynamic programming in Markov chains as in Or Goldfarb and Yoav Kislev, Management Rules for the Water Economy under Uncertainty, *The Economic Quarterly* (2002): 49 (2) 602–25 (Hebrew).

CHAPTER 11

Current Water Provision and Allocation in Palestine

ALFRED ABED RABBO

The main source of water for Palestinians is from rainfall stored in the aquifers of the Upper Cretaceous eastern basin flanking the central mountains, the northern Eocene basin, and the shallow coastal aquifer of Gaza (figure 11.1). Rainfall is limited to the winter months and, during 1998–2000, was considerably below the average. About 75% of the annual rainfall evaporates. Average annual groundwater renewal is on the order of 630 Mm^3 for the West Bank and 42 Mm^3 in Gaza, of which Palestinians receive about 110 Mm^3 from the West Bank aquifers and less than 20 Mm^3 from the Gaza aquifers, roughly 20% of the total annual renewal. About 80% of these waters are exploited by Israel.

Palestinians will continue to live under conditions of significant water stress for the foreseeable future. According to World Bank[1] estimates, the gap between demand and supply is –32% and will be –55% in 2020. All fresh water in the region should be reserved for domestic use, with treated wastewater supplying agriculture and industry. Half a century of mismanagement, including draining wetlands and over-pumping of aquifers, has reduced the quantity and quality of available water resources[2] in the area. It is to be hoped that the final status agreements reached in the bilateral peace talks between Israeli and Palestinian negotiators might lead to a more responsible shared management of these scarce water resources.

The only operating wastewater treatment plant in the West Bank is in El Bireh, near the town of Ramallah. Part of the current West Bank Integrated Water Resources Program III[3] involves the construction of a wastewater treatment plant to the south of Hebron. Treated wastewater is an essential resource for non-domestic uses.

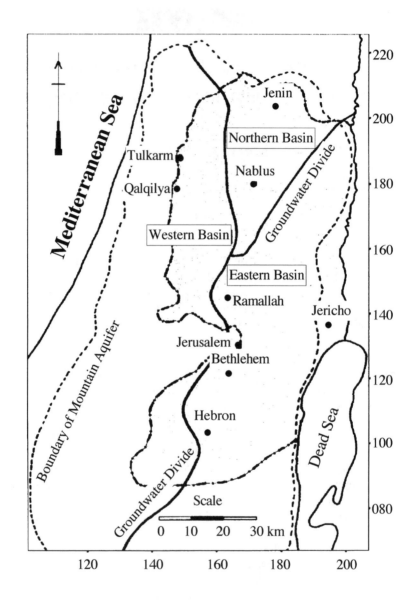

Figure 11.1 West Bank Aquifers.

Water Allocation for Agriculture

The agricultural sector consumes 70% of the water resources available to Palestinians in the West Bank even though irrigated agriculture represents only 5% of the total land available for farming. Israeli settlers consume six times more water for irrigation than Palestinians in the West Bank. Less than one percent of the land available for agriculture is irrigated in the southern West Bank. Palestinian agriculture in the West Bank consumes 84 Mm^3 of water.[4]

Southern West Bank

Water supply for Palestinians in the southern West Bank comes from a variety of sources. Parts of some of the larger towns and villages are connected to the Israeli network, receiving water from the deep wells in the Herodion well field (figure 11.2). The aquifer potential of the southern part of the eastern basin of the Mountain Aquifer was recognized by Rofe and Raffety[5] and the Beit Fajjar 1 was drilled whose waters supply East Jerusalem and the Bethlehem and Hebron districts. The Israeli water company, Mekorot, drilled five deep wells, Herodion 1 – 5, between 1971 and 1993. The Beit Fajjar well was deepened in 1988. These wells have provided high quality drinking water to the indigenous Palestinian population and the Israeli settlers for more than a quarter of a century. The sustainability of this well field is in doubt. Data from the Israel Hydrology Service (IHS) over this period shows an average drop in the water table of up to three meters per year.[6,7]

In accordance with the requirements of Article 40 of the Interim Accord,[8] whereby an extra 51.4 Mm^3 annually would become available to the Palestinian population, two wells, Hebron 1 and 2, were drilled in 1996 by Mekorot for the Palestinians. These became operational in 1999, supplying water to Hebron. Through a two-stage well development program undertaken by CDM Morganti, an American engineering consortium, contracted by the Palestinian Water Authority (PWA), sixteen new production wells were drilled in the eastern basin of the Mountain Aquifer (1999–2001).

Engineers working for the PWA on the new wells estimate a further 100 m drop in the water table over the next quarter of a century.[9] Aliewi and Jarrar[10] estimate the sustainability of this part of the aquifer as being far shorter, on the assumption that the eighteen new wells in the southern part of the Eastern Basin of the Mountain Aquifer will be pumping at the projected rate of 250 m^3/hr, while the seven older wells will continue to pump at their present rates. The critical report of de Bruijne et al.[11] blames

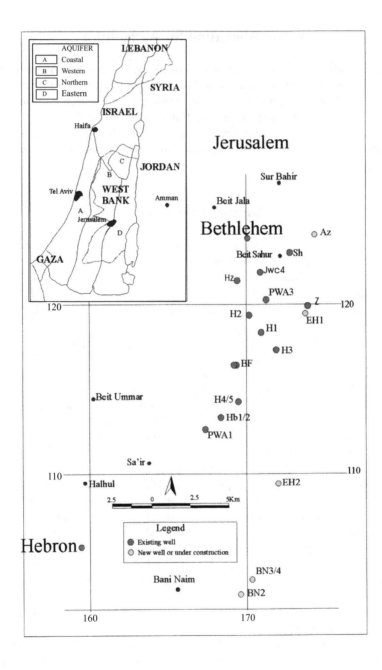

Figure 11.2 The Herodion Well Field.

the European and American donors for setting up an unsustainable exploitation of the aquifer. Clearly the PWA, now in a position to supply the long-deprived Palestinian population with an abundance of water, did not have conservation or even sustainability as their major priority.

The Water and Soil Environmental Research Unit at Bethlehem University (WSERU) noted bacterial contamination of water from three of the Mekorot wells[12] and in four of the new PWA wells during pump testing. Karstic aquifer drainage allows rapid flow from the surface to below the water table, permitting colonial growth of coliform bacteria at depths around 250–280 m. Most of the wells penetrate the clay-marl seal separating the unconfined sub-aquifer from the confined sub-aquifer and reach depths of between 700 and 800 meters. Should all of the Mekorot and PWA wells pump at full capacity the sustainability of the well field would be seriously threatened. Politicians and aid agencies deny that this is the case. Exploitation of the confined sub-aquifer will be considerably more expensive than that of the phreatic sub-aquifer. Conservation and sustainability, despite protests to the contrary, are not treated as a priority by those political and engineering agencies engaged in exploiting the aquifer.

Dependence on springs for domestic water and the collection of rainwater in household cisterns, augmented by expensive tanker supplies, are the main sources of water for many of the villages, particularly to the west of the hydrologic divide in the southern West Bank.[13] Private operators fill tankers from their own springs or buy well water and sell it to villagers at a much higher price than that charged for network water. There are no controls on these private operators. Most springs and cisterns are contaminated with bacteria. A few of the smaller, more remote villages, such as 'Imreish, Assura and Marah al Baqqar, to the south of Dura, are almost completely dependent on these tankers. Since 1999 Dura has received water from the upgraded Fawwar well located 5 km to the east of the town. However this supply is controlled by the Hebron Municipality and very often is diverted to Hebron City at the expense of Dura.

BF	Beit Fajjar 3 (Mekorot)
H	Herodion wells 1-5 (Mekorot)
Jwc4	Jerusalem 4 (IDF Camp well)
Hb	Hebron 1 & 2 (Mekorot)
Hz	Hundaza (PWA)
PWA	Palestinian Water Authority wells 1,3 &11
Z	Zatara (PWA)
EH	Extra Herodion 1 & 2 (PWA)
BN	Bani Naim 2 & 3 (PWA)
Sh	Shdaimah (PWA)
Az	Azzariya (PWA)

Within the villages there are conflicting views as to the prioritization of water use. The major concern of housewives is for a sufficient supply of domestic water, particularly of safe drinking water. In some villages provision of irrigation water takes priority at the expense of domestic water. Usually the women do not have a strong voice, and village councils are completely dominated by men.

Provision of Domestic Water Supplies in the Bethlehem and Hebron Districts

The first priority for the PWA in water allocation is the provision of drinking water and other domestic uses. The stated water policy of the PWA is that all Palestinians have access to drinking water of a quantity and quality at least up to the standards set by the World Health Organisation (WHO).[14] The use of groundwater for other than domestic purposes is to be reduced. The use of alternative sources of water (e.g. treated wastewater, surface water, runoff, rainwater harvesting) is to be used for non-domestic purposes, and these are to be enhanced.[15] Water resources, including surface and ground waters, are to be protected from over-extraction and from all sources of pollution.[15] Water resources are, in the short-term, scarce. The official line of the PA, however, is that in the long term, provided that Palestinians have access to their water rights, there will be a surplus in the West Bank Governorates.[15] This viewpoint assumes that the present Israeli settlements in the West Bank will be evacuated and that only Palestinians will have access to the West Bank aquifers.

In the Bethlehem Governorate, 89.8% of households have access to tap water, and in the Hebron Governorate it is 66.3% (table 11.1). Of these, 31% in the former and 23% in the latter experience cutoffs in the water supply (table 11.2). Only 10% of households in the Bethlehem Governorate and 12.8% in the Hebron Governorate are connected to a public sewage system (table 11.3).[4]

Table 11.1 Households by main source of drinking/cooking water

Main source of drinking/cooking water	% Bethlehem Governorate	% Hebron Governorate
Indoor public system	89.8	66.3
Outdoor public system	3.6	4.8
Individual well	4.6	24.1
Tanker truck	0.2	2.8
Other source	1.9	2.1

Source: PCBS, 1996.

Table 11.2   Households by frequency of use of secondary source of drinking/cooking water

Use of secondary source of drinking/cooking water	% Bethlehem Governorate	% Hebron Governorate
Once a week	31.0	23.0
Once a month	12.1	8.6
Rarely	13.0	20.7
Never	44.0	47.7

Source: PCBS, 1996.

Table 11.3   Households by type of sewage disposal system

Type of sewage disposal system	% Bethlehem Governorate	% Hebron Governorate
Public sewer	10.0	12.8
Septic tank/cesspool	85.2	80.7
Other means	4.8	6.5

Source: PCBS, 1996.

Overall, only 35% of Palestinian households are connected to a public sewer system; therefore raw sewage flows from cesspits into wadis (dry river valleys) and, in some cases, because of the karstic nature of the aquifers, into the groundwater systems. There are also quite inadequate facilities for solid waste disposal, another potential source of pollution of the aquifers.

Although 85.6% of households have piped water, only 35% are attached to a piped sewage network. Over 120 villages in the West Bank do not have access to drinking water networks.[16]

The Northern West Bank

There are two main aquifer systems in the northern West Bank. The Eocene aquifer is contained within a shallow synclinal structure of the flanking and underlying Upper Cretaceous strata. Groundwater flows in a north-easterly direction in the Eocene basin. As well as deep production wells for domestic uses, there are many hundreds of relatively shallow private wells mainly for agricultural purposes. The eastern and western aquifer systems, within the Upper Cretaceous, in places superficially covered by Quaternary deposits, drain towards the Jordan Valley and Mediterranean Sea respectively. The Eocene system depends directly on the renewable recharge from precipitation, receiving an average annual rainfall of about 500 mm. In the plains around Tulkarm, Pleistocene for-mations of alluvial deposits crop out and form an additional shallow aquifer system. Direct recharge to the aquifer systems takes place over the entire area, with an average annual precipitation of about 600 mm; how-ever, discharge rates remain constant even in the dry season, indicating an abundant reservoir source of water.

Pollution

Detailed chemical analysis of samples collected by WSERU from the shallow wells of two unconfined aquifer systems in the northern West Bank revealed serious levels of pollution. There is a risk of potential health hazards because of a lack of environmental control. It is important that the Palestinian Water Authority effectively apply well protection policies.

The major sources of groundwater pollution in the northern part of the West Bank are ill-considered agricultural activities and careless wastewater disposal. Pollution due to agricultural activities is caused by an excessive use of fertilizer, coupled with over irrigation, facilitating passage though the unsaturated zone to the groundwater aquifer. Farmers in the West Bank use chemical fertilizers to improve their crops. The most commonly used fertilizers are ammonium sulphate, urea, potassium nitrate, and super phosphate. Therefore, the most important ions added to the recharge areas of the shallow aquifer are nitrate, ammonium, potassium, sulphate, and phosphate. The pollution due to these agricultural activities may appear in the form of increasing levels of electrical conductivity (EC) and nitrate. In some cases, high concentrations of potassium and sulphate are recorded. The concentration of potassium in the groundwater is normally low. This is because most of the potassium is absorbed by plants or adsorbed by mineral particles, particularly clay minerals, in the soil. Clear directives must be given to farmers concerning the safe application of fertilizers.

Uncontrolled wastewater disposal sometimes contaminates other waters causing an increase, particularly in EC values and concentrations of chlo-ride, sodium, nitrate, and sulphate. In those areas not served by sewage systems, wastewater from septic tanks can pollute the shallow aquifer systems. In those served by sewage systems, there may be leakage from the sewage network or from poorly sealed wastewater collection pools.

Biological contamination is common among the shallow wells in the northern West Bank. A few of the wells were found to be chemically unsatisfactory for drinking water purposes. It is important that these shal-low wells be protected and rehabilitated, where possible, and proper storage facilities provided. This would facilitate upgrading the water to good potable standards by disinfection and other appropriate methods of treatment. Legislation preventing sewage disposal into wadis would provide some protection for the aquifer from this form of pollution. Cesspits are common, particularly in the rural villages, and present a pollution danger to the aquifer and spring discharge. Strict regulations requiring proper seals to cesspits or their replacement where possible, with proper sewage networks, could remove this danger. Unregulated use of fertilizers can be a serious source of contamination. This, together with excessive irrigation, a combination observed in some agricultural areas in the northern West Bank, is a source of contamination that could be avoided with proper legislation.

Gaza

The Gaza Strip is 365 km^2 in area with a population of about 1 million Palestinians and 6,600 Israeli settlers.[4] There continues to be a decline in water quality from the shallow coastal aquifers that are located in an interfingering complex of sands and sandstones separated by impermeable clay seals.

Direct rainwater infiltration is about 40 Mm^3/yr while underground flow from the Mountain Aquifer can be 10–20 Mm^3/yr. However, a series of Israeli wells to the east of the Gaza border extract a considerable amount of this westward flowing groundwater. Excessive exploitation of the delicate coastal aquifer is unsustainable. Domestic consumption is about 60 liters daily per capita (WB = 88, Israel = 350, Jordan =1531, Egypt =3371).[4] It is therefore understandable that as much water as possible is extracted from the aquifers, legally and illegally.

Pollution comes from the surface: from sewage flows and cesspits; from agricultural wastes, pesticides, and fertilizers; from sea-water intrusion as a result of unregulated drilling and consequent pressure releases; and from saline waters located under the coastal aquifers, again, rising as a result of pressure release.

Agricultural use of water is inappropriate but continues because of livelihood, cash cropping, and food security. Citrus is an extremely water-expensive crop but remains an important cash crop for local and foreign markets as do other water-demanding crops. Wastewater treatment provides an increasing amount of irrigation water.

Since the PWA has taken responsibility in Gaza, losses through leakage from pipes have declined considerably. It seems unlikely, though, that the Gaza coastal aquifer can be saved even if the necessary, but very difficult, political decisions concerning water prioritization and allocation can be put into effect. Some form of desalination would seem to be essential for providing good quality drinking water.

Conclusion

The West Bank and Gaza have a low adaptive capacity that has a serious negative impact on its problem of major water insecurity. It is within the context of a fragile economy, dependent in large part on foreign aid, and a weak government striving for a statehood that can have only very limited sovereignty, that fundamental decisions concerning the diminishing water resources will have to be made. Agriculture is not only an important source of livelihood; it is, for many villages, an essential provider of food. Diversification is therefore not an easy option.

Some alleviation of the water stress can be obtained by developing waste-water treatment plants to provide for non-domestic water use and desalination, particularly in Gaza, for drinking water purposes. It is to be hoped that the final status agreement will allow for a more sustainable exploitation of the West Bank aquifers.

Notes

1   Palestinian Economic Council For Development and Reconstruction (PECDAR), *Palestinian Water Strategic Planning Study* (Ramallah: Palestinian National Authority, 2001).
2   Ibid.
3   USAID/PWA West Bank Integrated Water Resources Program III, Project Kickoff/Partnering Meeting, September 21, 2000, Hebron (2000).
4   Palestinian Academic Society for the Study of International Affairs, *PASSIA Diary 2001* (Jerusalem: PASSIA Publication, 2001).
5   Rofe and Raffety, Consulting Engineers, *Jerusalem and District Water Supply, Geological and Hydrological Report* (London: Ministry of Water and Irrigation, the Hashemite Kingdom of Jordan, 1963).
6   Israel Hydrology Service, *The Hydrological Year Book 1977* (Jerusalem: Ministry of Agriculture, Water Commissioner, 1978).
7   Israel Hydrology Service, *The Hydrological Year Book 1989* (Jerusalem: Ministry of Agriculture, Water Commissioner, 1990).
8   Israel–Palestinian Bilateral Negotiating Team, *Water Supply and Sewage Disposal* (Washington: Interim Accords, Article 40, 1995).
9   Henning Moe (personal communication), Chief Engineer, Camp Dresser and McKee International Inc. (CDM) 1999.
10 A. Aliewi and A. Jarrar, Technical Assessment of the Potentiality of the Herodian Well Field Against Additional Well Development Programmes (Ramallah: Palestinian Water Authority, 2000).
11 G. De Bruijne, J. Moorehead, and W. Odeh, Water for Palestine: a Critical Assessment of the European Investment Bank's Lending Strategy in the Rehabilitation of Water Resources in the Southern West Bank, Brussels: *Reform the World Bank Campaign Report*, Palestinian Hydrology Group, 2000.
12 A. Abed Rabbo, D. J. Scarpa, and Z. Qannam, A Study of the Water Quality

and Hydrochemistry of the Herodion–Beit Fajjar Wells, *Bethlehem University Journal* 17 (1998): 11–28.

13 D. J. Scarpa, The quality and sustainability of the water resources available to Arab villages to the west of the divide in the southern West Bank, *Water Science and Technology* 42, 1/2 (2000): 331–36.

14 World Health Organisation (WHO), *Guidelines for drinking-water quality* (New York: WHO, 1995).

15 Ministry of Planning and International Cooperation (MOPIC), *Valuable Agricultural Land in the West Bank Governorates* (Ramallah: Palestinian National Authority, 1998).

16 Palestinian Central Bureau of Statistics (PCBS) *The Demographic Survey in the West Bank and Gaza Strip* (Ramallah: Palestinian National Authority, 1996).

CHAPTER 12

The Peace Process and Water Supply in Jordan: Inter- and Trans-Boundary Border Projects

MOHAMMED ABUDAYYEH MATOUQ

In the 1990s, Jordan entered into direct negotiations with Israel with the goal of reaching an agreement that would allocate the Jordan River resources in a way that would provide it with good drinking water, particularly for the city of Amman. On Febuary 13, 1996, following the Israeli-Jordanian peace agreement signed in 1994, Israel, Jordan, and the Palestinians set forth in Oslo, Norway a Declaration on Principles for Cooperation on Water-Related Matters as part of the program adopted by the Working Group on Water Resources that was established in Madrid in 1991. This Declaration defined common denominators, and principles for cooperation, for new as well as augmented water resources, and specified the proposed areas for cooperation at the regional level to manage water resources and allocations.

The Bilateral Trans-Boundary Upper Jordan River Water Resources Management Scheme between Jordan and Israel

Following their peace agreement, Jordan and Israel set down a strategy for allocating water resources (mainly of the upper Jordan River), for exploring additional water resources, and for seeking out a management policy framework. The outcome of their discussions was the establishment of a water working group that played a significant role in determining the details of the water annex to the peace agreement and in proposing the projects to be implemented, i.e., those dealing with water resource allocations and development of water resources in both countries. These proposed projects are classified under three main action categories: (1)

pipeline construction; (2) building dams; and (3) securing new additional water resources such as desalination plants.

In Annex II (Water-Related Matters) of the Jordanian–Israeli Peace Agreement, it is stated in Article 1 that Jordan and Israel agree to allocate and distribute the waters of the Yarmouk River among the countries as follows: During summer each year (15 May to 15 October), Israel will pump 12 Mm^3 from the Yarmouk River with the remainder going to Jordan. During winter each year (16 October to 14 May), Israel will pump 13 Mm^3 and Jordan is entitled to the remainder. Also during winter an additional 20 Mm^3 can be pumped by Israel from the Yarmouk River to Lake Tiberias and in return Israel will supply to Jordan the same amount from Lake Tiberias during summer (see below).[1]

The two countries agreed on similar terms for sharing the waters of the Jordan River. During summer Israel agreed to transfer 20 Mm^3 to Jordan from the Jordan River directly upstream of the river and from the Degania gate (i.e., Lake Tiberias) in return for a similar amount to be pumped by Israel from the Yarmouk River during winter. It should be noted that this water supply is like a water-loan, with winter Yarmouk River water being temporarily stored in Lake Tiberias to be returned to Jordan during the summer (as stated above). All costs associated with transferring water to Jordan are to be covered by Jordan. The two countries also agreed to share the waters of the Jordan River downstream of its confluence with the Yarmouk River. During winter, Jordan is allowed to store for its use a minimum average of 20 Mm^3 upon building a storage reservoir on the river to accommodate increased flooding. Israel has the right to use 3 Mm^3 per year of this stored amount when its original storage capacity is exceeded. Additionally, both countries are entitled to equal shares of Jordan River water between its confluence with the Yarmouk and its confluence with Wadi Yabis such that Israel would maintain its current usage of those waters and Jordan would be entitled to the same annual quantity. Finally it was stipulated that Jordan's supply should not harm Israel's supply.

Currently, Israel diverts approximately 20 Mm^3 of saline waters from the Jordan River basin around Lake Tiberias and puts it back into the Jordan River south of the lake. Jordan is entitled to an annual quantity of 10 Mm^3 of water from the desalination of those 20 Mm^3. As the desalination facility is not yet operational, Israel agreed to supply an equivalent amount of water (i.e., 10 Mm^3) directly from Lake Tiberias (but outside of the summer period) until the desalinated water could be supplied. Israel also agreed to explore the possibility of financing the operation and maintenance cost of the supply to Jordan of this desalinated water (not including capital cost).

In addition to sharing the waters of the Yarmouk and Jordan Rivers, Israel and Jordan agreed to cooperate in finding sources that would supply

Jordan with an additional quantity of 50 Mm^3 annually of water of a drinkable standard.

In Article 2 ("Storage") of Annex II to the peace agreement, it is stated that the two parties will jointly attempt to develop a storage facility on the Yarmouk River, thereby securing a new water resource. Indirectly, this will undermine the rationale for the Al-Wahda ("Unity") Dam. The idea for the Al-Wahda Dam emerged following opposition to the Johnston Plan by Israel in 1956. At that time, Jordan and Syria, with support from the Arab League, agreed to cooperate in building a dam on the upper stream of the Yarmouk River. This deal would have allowed Syria to receive 75% of the total generated hydroelectric power with Jordan being the beneficiary of the water. The dam at that time was estimated to be able to supply Amman with 50 Mm^3 per year. However, since the dam would have been built on the upper stream of the Yarmouk River, the downstream supply of water into Israeli territory would have been reduced. Therefore, Israel was completely opposed to this project.

After the peace agreement, the Al-Wahda project was re-initiated but with a different proposed location. As the peace agreement states: "*both parties – Jordanian and Israeli – agree to cooperate in the building of a diversion dam on the Yarmouk River directly downstream of the existing diversion point,*" known as Adassiya. This means that the dam is to be built on Jordanian and Israeli-controlled land. However, when the announcement was made by both sides to start initiating the dam project in August 1997, the Syrians opposed the idea, which caused a political conflict to develop between Jordan and Syria. This was for two reasons: (a) the proposed dam would be located on disputed land (the Golan Heights) that had not yet been settled between Syria and Israel; and (b) the project went against what Jordan and Syria had agreed upon in 1956, viz. to build the dam on the upper stream of the Yarmouk River,[2] with Syria receiving the generated electricity.

The proposed dam would have an 8 Mm^3 storage capacity and would be able to yield approximately 14 Mm^3 annually. This stored water from the dam would be conveyed to the King Abdallah Canal (KAC) also known as the East Ghor Main Canal. The project is estimated to take about three years for completion. However, the dam project is still under negotiation with respect to when and how to implement it, as it now requires a multilateral agreement between Jordan, Israel, and Syria.

Water Supply Projects for Amman

Jordan's main water scarcity is in drinking water. This is especially so in the city of Amman, and the government of Jordan has focused most of its proposed projects on supplying water to Amman and its greater munici-

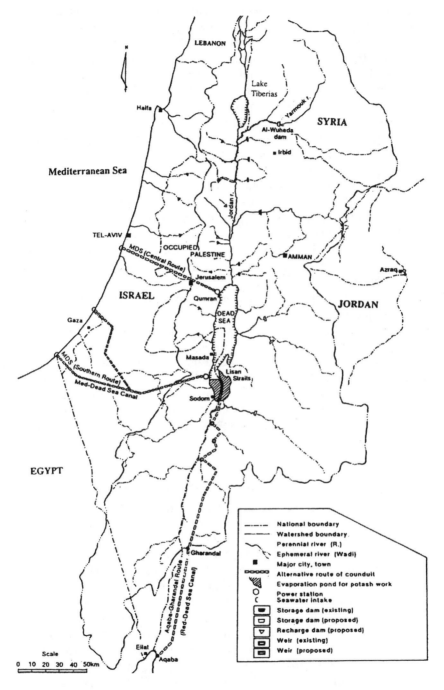

Figure 12.1 The Current Water Distribution System of the Jordan River.

pality. (Hereafter Amman refers to the city and its greater municipality.)

The existing Deir Alla–Amman Pipeline supplies 110 Mm³ of water from the KAC to Amman and contributes about 14% of the total water resources of the country. The KAC is the main water conveyor for Jordan in the Rift Valley and originates at the Yarmouk River and terminates near the Dead Sea. Additional water resources from side wadis (seasonal streams) are diverted into the canal. This water is primarily designated for irrigation. Taking advantage of the existing KAC–Deir Alla–Amman pipeline, and in compliance with the Peace Treaty, two projects have been proposed to develop the water resources on the Jordan River by building additional pipelines (figure 12.1).

First, a new pipeline that includes five pumping stations to draw water from the Deir Alla reservoir to Amman is to be constructed from Deir Alla to Amman with dimensions of approximately 38 km in length and 180 cm in diameter. However, since the Yarmouk River's main source is rainfall runoff, the shortage in water supply during the dry months will be significant, and the project has to realize its potential annual capacity in order to be feasible.

Second, a new proposed pipeline of length 60 km and diameter of 70 cm is to be installed inside Jordanian territories, parallel to the KAC. It would supply Amman with 50 Mm³ of desalinated water yearly via the existing Deir Alla–Amman pipeline. This portion of desalinated water would be supplied following the construction of a desalination plant to be located south of the Sheikh Hussein Bridge. Water for desalination will be allocated from the Beit She'an Valley saline spring water in Israel and will include the diverted saline water from Lake Tiberias discussed above. The project will cost approximately US$350 million according to an Israeli feasibility study.

Analysis of the Peace Projects and Future Prospects If the Peace Agreement Is Terminated

The peace agreement between Israel and Jordan included a package of water projects to be implemented along both the West and East banks of the Jordan River. It should be noted, though, that the peace agreement does not include any proposal for water resources allocation inside the West Bank other than describing how to develop the existing conventional water resources. None of these projects, however, included details of water allocation from Jordan's upper stream to Israel. In effect, this has meant that water allocations from the Jordanian side have been suspended. Therefore, most of the proposed projects in the peace agreement are within Jordanian territories and allow Jordan to focus its efforts on developing its internal resources in a number of ways: by building dams and/or storage reservoirs on the Yarmouk River before its confluence with the Jordan River such as the Addasiya dam; and by extending pipelines, namely the existing KAC and the proposed KAC–Deir Alla–Amman pipelines. As yet, the proposed pipelines have yet to prove their feasibility because of the irregular rainfall amounts into the Yarmouk River.

The desalination plant that has been proposed on the Israeli side near Beit She'an, which would supply Jordan with drinking water, might in the long term be subject to political dimensions, and water flow might be affected if any political conflict takes place.

Within the agreed-upon framework of the projects in the peace treaty, Jordan was to benefit by receiving 200 Mm^3 of water per year, based on two strategies: (a) efficient water resources development inside the country, such as looking for new ground water resources, water harvest, and more re-use water policies; and (b) additional water resources from the Israeli side, either from the Jordan River basin or new water resources. However, up until now, Jordan has received less than half of its allocated amount agreed upon in the peace agreement. Shortly after the peace agreement was signed, during 1994–1997, only 50 Mm^3 per year of water had been supplied to Jordan from the Jordan River. This reluctance and delay from the Israeli side in fulfilling its commitment led the late King Hussein of Jordan to meet with the Israeli prime minister in September 1997 to urge the Government of Israel to abide by its promises and supply Jordan with the additional portion of water. This intervention by the King resulted in an agreement to provide another 25 Mm^3 per year. Though the Israeli government agreed to the King's request, no clarification was forthcoming about where this portion would come from.[3] This lack of clarity in managing water issues means, in effect, that they may be raised again whenever any conflict arises or when there is a shortage in rainfall in the region that would result in a high demand for water supplies in Jordan and Israel.

These peace-related projects will be a financial burden to Jordan's government since they will impose new budget demands for the development of its internal water resources by building dams, desalination plants, sewage water treatment, new pipelines, and storage reservoirs. The proposed projects require more than five suggested storage reservoirs and three proposed dams. It is necessary to bear in mind the great economic impact of the Gulf wars on Jordan and how they have effected the economic situation in the country and continue to do so. These unpleasant economic realities, as well as uncertainty in the costs of the projects, will no doubt foster political disagreements that will lead to delays in their implementation.

According to the peace agreement, Jordan is projected to increase its water resources to 300 Mm^3 per year by the year 2010, 65% of which would come from developing internal water resources; Israel would maintain its current usage of the Jordan River basin while adding new resources that would come from ground water, desalination, and reused water. The peace agreement sets forth a principle for water cooperation and for the improvement of water-demand management; surprisingly, however, the agreement did not mention water sharing or water allocation of the Jordan River.

An important outcome of this agreement was a regional declaration for developing new water resources at the regional level. This unprecedented institutional initiative, the Middle East Desalination Research Center (MEDRC), was officially established in December 1996 during the negotiations held in Oman between Jordan, Israel, and the Palestinian Authority (PA). The founding members of this Center (located in Muscat, Oman) were Israel, the US, Japan, Korea, and Oman. The Center will be responsible for implementing the proper technology in the region to utilize sea water to increase water resources at the regional level.

The agreement does not articulate any means to deal with political conflicts that might have an effect on water allocation. As we have seen above, such potential conflicts could well have an impact on the Al-Wahda Dam project, which still needs clarification from Syria, and the proposed desalination plant that is to be located on the Israeli side. Even political tensions could make it impossible for Jordan to meet its water supply deficit. For example, a political conflict between Jordan and Israel could affect the water supply from Addasiya Dam intake, stop the water supply from Lake Tiberias, or interfere with the proposed supply of desalinated water. Therefore, Jordan must take into consideration the need to seek new water resources if for any reason the peace agreement is terminated due to unexpected political issues.

Due to such potential political conflicts, several scenarios should be taken into consideration. Jordan might acquire new water resources from neighboring countries. For example, it might enter into an agreement with Iraq to draw water from the Euphrates River through pipelines to Amman by connecting with the existing Azraq pipelines in the Eastern Desert region near Jordan's border with Iraq (see figure 12.2). Another proposal is to enter direct negotiations with Syria and Israel to convince the two parties to build the Al-Wahda dam. This is especially crucial inasmuch as Syria has begun implementing a small and medium-sized reservoir to reduce the flooded water into the Yarmouk River. This has created an obstacle to proceeding with the Adassiya Dam project. The lack of trilateral negotiations between Israel, Jordan, and Syria will work against Jordan's long-term interests since the Adassiya Dam will not be built if Syria is against this proposal. If Jordan and Israel were to start building the Adassiya Dam, Syria could build small dams on the upper streams to prevent water flow into the Yarmouk River, thereby rendering the proposed Adassiya Dam unfeasible. It should thus be clear that the three parties need to cooperate in building the Al-Wahda Dam, regardless of its location, if Jordan is to receive additional water from the Yarmouk River.

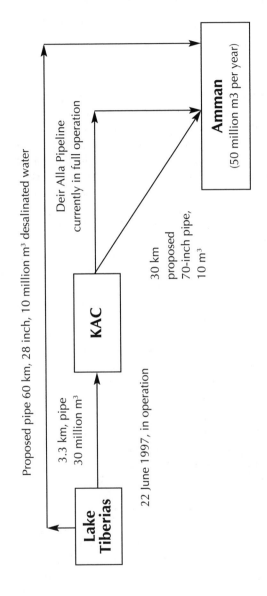

Figure 12.2 Water Allocation to Amman after the Peace Agreement.

Another possibility Jordan might well consider to offset the results of potential future conflicts would be to build a desalination plant in the Gulf of Aqaba without further delay to utilize the Red Sea water as a new water resource.

Conclusion

The water scarcity issue has played an important role in the implementation of a peace agreement in the region. The benefits of the resulting water agreement, however, have yet to be realized due to the critical political conflicts within the region. Jordan, which shares the longest border with Israel and has the longest frontage on the Jordan River, has realized that peace is the most reliable way to assure its water resources. Through the peace process, Jordan will be able to allocate a new 50 Mm^3 from Lake Tiberias and from the ongoing project to be built on the Yarmouk River.

Jordan's main water resources are from underground and surface water. However, underground water can safely supply the country with only 277 Mm^3, while the demand is 515 Mm^3.[4] Other surface water resources will not be able to satisfy this demand. Entering into a peace agreement will give Jordan the opportunity to increase surface water resources by the better management of Jordan River resources and the more efficient use and management of internal water resources.

There is no doubt that Israel and Jordan shall seek to increase water resources in the region by proposing major regional projects in the next decade, such as the Red Sea–Dead Sea canal, the Mediterranean Sea–Dead Sea canal, and desalination projects. These projects need regional collaboration and investment to secure water resources; this has been made possible by the peace agreement. After a 40-year history of war, the treaty with Israel has, for the first time, opened new opportunities for regional water resources development through the establishment of MEDRC, an example of an extra-regional development initiative. The management of the Center is Israel's responsibility, financial support is shared among the Arabs (Oman), Europe, the US, Israel, Japan, and Korea, and the working force will be from the region (figure 12.3).

However, just three years following the signing of the peace agreement, the Israeli side found it difficult to concede 100 Mm^3 per year of water without decreasing their supplies from Lake Tiberias. The very difficult issue will become even more unwieldy and complicated once the West Bank's water-share is included in the sharing arrangement. Thus the parties need to go beyond what they have agreed to thus far in order to maintain the spirit of the peace.

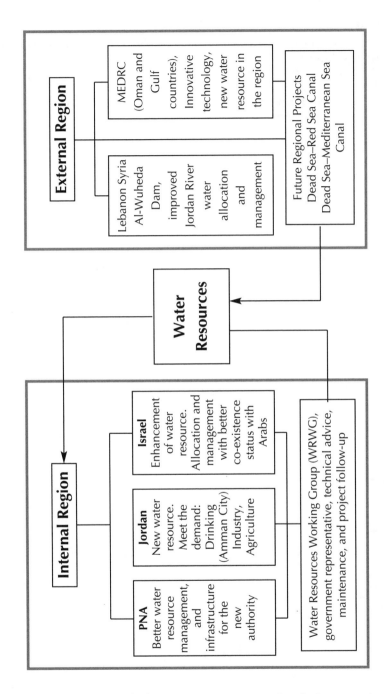

Figure 12.3 Current and after Peace Process Water Supply System to Amman City.

Notes

1   Israel–Jordan Peace Treaty Annex II, 26 October 1994, Ministry of Foreign
    Affairs of Israel.
2   *Star,* Amman, 28 August 1997.
3   *The Economist*, 17 May 1997: 52.
4   JICA, *"Country Study for Japan ODA to the Hashemite Kingdom of Jordan"*
    (Tokyo: March 1996).

CHAPTER 13

An Economic Approach for Making the Most of Jordan's Water

MOHAMMED ISSA TAHA ALI

Water resources in Jordan have considerably decreased since 1980. Irregular rainfalls, strong evaporation, and factors of strong population growth worsen the water deficiency. These facts have been emphasized by many studies that examined Jordan's water problems in the last 15 years.[1] These studies have also emphasized that Jordan does not have enough water supply for the future and should more efficiently manage its meager water resources. Other studies emphasized the role of water as a key factor in creating and sustaining peace in the Middle East.[2]

Jordan's economic growth rates have been low. From 1988 to 2000, annual GDP growth rates were, on average, 3.4%.[3] The poor management of water and the fluctuations of rainfall have resulted in a decreasing agricultural production per capita and a decline in per capita consumption of water. The growth rates of the agricultural sector were negative, at –2.6%, during the same period.[4] From the socioeconomic point of view, Jordan faces unemployment, poverty, and low productivity in the agricultural sector. For Jordan, economic performance is highly tied, among other things, with the status of water. As a developing economy, a large proportion of the labor force is engaged in agriculture that consumes most of the country's water supply. All these factors make it necessary to search for solutions for this nagging water problem. Examining the available options and the experience of Jordan's management of its water problem, I suggest an economic approach to alleviate the problem or at least a major part of it.

Status of the Water Sector in Jordan

During the late 1990s, a water crisis hit not only the population of Jordan but also the population of the Jordan River basin, notably the Palestinians and the Israelis. The winters did not bring the expected loads of rain that normally refill the Jordan River, the Sea of Galilee, and the underground natural water storages under the West Bank.[5] By the end of winter of 2001, the water collected in King Talal Dam amounted to 11.5 Mm³, Wadi Al Arab had 8.2 Mm³, Kafrein increased its storage to 1.4 Mm³, Shurahbeel Dam contained 1.7 Mm³, Karama Dam stored some 1.5 Mm³, and Shuaib Dam contained 1.4 Mm³.[6]

To cope with the threatening water scarcity each summer, a rigorous water-rationing schedule has been put in place according to which households are supplied with water once or twice a week. Also farmers in the Jordan Valley, who completely depend on irrigation for their crops, normally receive water twice a week for 12 or 18 hours depending on water pressure prevailing in the respective area, size of the farm unit, and the crops planted. While tentative water rationing schedules for the summer are altered every month according to rainfall received and depending on estimates of water availability during summer, the rationing plan for cities generally remains unchanged because drinking water supply to municipalities has priority.

The average loss from Amman's water network is 52%, higher than Tunis – 21%, Gaza – 31%, and Casablanca – 34%. It is comparable to Sana'a – 50%, and Algiers – 51%. Among Arab capitals, only Damascus with 64% manages to lose more water. Even though there is some leakage into the ground (because part of the network is old and deteriorating) and some water leaks onto the streets, it is believed that a major part of the water is actually lost because of illegal extraction of water. Some illegal connections to the network are reputed to be known to the Water Authority but can only be stopped if the Water Authority dares to face the offenders.[7]

A major economic problem in the Jordanian water sector is the failure to recover the costs of operation and maintenance. The Jordanian Water Authority is expected to lose some JD50 million ($70 million) a year, which calls for either higher prices or improved efficiency of network management, or both. Some 6% of the labor force is directly employed in agriculture, either as farmers or farm workers, while the agriculture-related service sector constitutes 28%. For 20% of the 5.4 million inhabitants of Jordan, agriculture is the main source of income.

Water Consumption

Jordan's daily per capita consumption of water is relatively low – it is estimated between 70 and 75 liters per day, far below the normal 200 liters per
day. It is the rapidly growing municipal sector and its ever-growing water
demand that necessitates more rigorous policies to manage water
resources. Besides convincing citizens of the necessity to conserve water
because of the country's dire outlook with regard to its diminishing
resources, economic benefits are believed to be highly persuasive. It is
simple and low-cost devices that can help a great deal with regard to water
conservation. Special devices for the showerheads and taps and different
toilet flushing systems could reduce domestic water consumption by
around 50%. Water saving devices for showerheads could reduce the water
flow rate from an average 20 liters per minute down to only 8.5 to nine liters
per minute. Saving devices attached to taps could reduce the flow rate from
12 down to six liters per minute. And modern toilet flushing systems could
function using only six instead of 12 liters as usually is the case in Jordan.

Shortages in water resources have reached crisis levels. The per capita
consumption of countries like Jordan has reached alarming levels.
Compared to what is internationally conceived as adequate water
consumption at 1000 cubic meters annually and water scarcity level at 500
cubic meters, Jordanians have a share of 350 cubic meters per capita. There
is no common vision between states in the region on how to manage water
crises especially in trans-boundary dimensions.

Per capita share of water consumption in the Jordanian capital is less
than 80 liters per day, which is rather low but looks reasonable compared
to Tunis – 80 liters, Gaza – 70 liters, Algiers – 70 liters, and Sana'a – 50
liters, although the amount of water that would be comfortable ranges
between 120 to 200 liters per person per day.

At another level we find that Jordanian staff, usually accused of laziness,
remain more efficient than others. The Jordanian Water Authority has 5.5
employees for each one thousand connections, against 6 in Casablanca, 7
in Gaza, 8.6 in Algiers, and 10 in each of Tunis and Sana'a.

In 2000, the winter's rainfall was near or above normal, whereas in 2001
it reached only 54% of the annual average. Although hotels, hospitals, and
large educational institutions form only 5% of Jordan's domestic water
consuming sector, they actually consume up to 80% of the annual domestic
water supply.

However, agriculture poses the biggest threat to water resources. The
sector consumes around 70% of the country's water while municipal and
industrial consumption reaches 26% and 4%, respectively.[8] In addition,
agricultural sector returns only 5% to the national economy. The gulf
between agricultural consumption and contribution to the GDP has led

economists and agriculture experts to advocate reducing agriculture's water allocation, not only in Jordan, but worldwide. Instead, the municipal and industrial sectors should be the top beneficiaries of water distribution.

Water Policies

The chronic water crisis in Jordan, and indeed in many developing countries, is a matter of imbalance between water supply and water demand. Jordan's water imbalance clearly portends a future crisis. The current annual water deficit of use (not true demand) is 321 Mm³. More water must be made available, and some uses need to be curtailed or modified to save water. Even if Jordan seeks to meet its increasing water demand by overdraft without attempting to reduce demand and enhance supply, the annual deficit could increase to nearly 1,200 Mm³ by the year 2015. Gains through supply increments and severe demand reductions suggested in this report would not hold the deficit to its present level. Other supply sources are scarce and expensive to exploit. Competition will increase among uses for the available supply. A number of alternative actions are possible that could help postpone a crisis. The options available deal with carefully increasing and managing the water supply and demand and protecting water quality.

Water policies are divided into two categories: those affecting water demand, and those affecting water supply.

Policies Affecting Water Demand

Reforming Water Subsidies

In 2001, Jordan Valley farmers paid only 15 fils (1000 fils per JD) per cubic meter of water on average. The real cost of providing water to farmers amounts to a multiple of that amount. From an economic point of view, these subsidies increase water use above what it would be if consumers had to pay the full costs. The need for water-pricing reform, especially for agricultural use, is high among the options that the Jordanian government should consider, an option that the government has avoided for political and social reasons. The government stresses that water costs more to supply than what farmers are charged. Therefore, it should look for production and supply schemes considering the actual price of water and not only the maintenance and operation costs. Substantial increases in water prices can be avoided if the efficiency of maintenance and operations can be increased, thereby reducing operation costs.

Privatizing Water Management

A consistent complaint of water conservationists has been that farmers plant high water consuming plants, such as bananas. If water management is privatized in the Jordan Valley, it is likely that the area will be planted with crops that consume much less water. As part of the privatization program, the Jordanian government has entered into a long-term contract with a specialized French company to manage the water network and distribution. Although such a step is an effective way to reform the water sector, the quality of the water utility, or lack of it, has remained more or less the same. Breakdowns continue. Water running down the streets is still a familiar sight.

Restricting Banana Planting in Southern Jordan Valley

Farmers in the southern Jordan Valley and Wadi Araba should be restricted to planting only five dunums (1 dunam = 1,000 m²) of bananas for each 30–40 dunam agricultural unit. The government has long sought to discourage the cultivation of high water-consuming crops. In the face of the country's chronic water shortage and nearly five years of drought, the government moved to regulate banana growing. Strict and decisive action will be taken against violators of this new regulation as well as other restrictions already in place. Violations include exceeding permissible water consumption limits and making use of state-owned agricultural lands. The daily water utilization per agricultural unit in the southern valley is limited to 259 cubic meters. Of the 46,000 agricultural dunums in the southern valley and Wadi Araba, some 5,000 dunums are planted with bananas which consume four to five times more water than other crops. Many agricultural experts believe it is preferable to import bananas rather than to plant them. It should be noted that it costs JD700 to cultivate one dunum (1,000 sq. meters) of bananas, although importing bananas costs JD630. The southern valley and Wadi Araba farmers traditionally cultivated primarily vegetables. But the increased scarcity of water in the northern Jordan Valley where most bananas were grown forced farmers there to opt for different crops, and since 1995 banana growing took root in the southern areas where water had been more available.[9]

Restructuring Water Authorities

The Ministry of Water and Irrigation (MWI) is responsible for the water resource allocation and water quality protection. MWI is composed of two authorities: the Water Authority of Jordan and the Jordan Valley Authority. The Water Authority of Jordan (WA) is responsible for the allocation of water to domestic, agricultural, and industrial users. The WA maintains environmental protection of surface and ground water quality from industrial and municipal wastewater discharges by monitoring effluent quality and taking compliance actions. The WA is responsible for enforcing effluent limits for discharges to surface waters including the Gulf of Aqaba. The WA is also responsible for the construction, operation, and maintenance of sewage treatment plants throughout the county. The Jordan Valley Authority is responsible for all irrigation water allocations in the Jordan Valley up to 500 meters in elevation. The JVA is also responsible for soil reclamation in the valley's fertile farmlands. This huge and fragmented list of public authorities, let alone NGOs and research centers, makes the environmental map in Jordan very complicated and multifaceted. This will eventually result in overlapping and sometimes competition by authorities.

To overcome such difficulties and to cope with the ever-increasing technicalities of environmental problems and issues, two very important improvements have to be implemented. The first includes raising the level of environmental decision to an independent and technical ministerial level instead of bureaucratically dependent levels. The second deals with utilizing high-caliber human resources and modernizing the structure of environmental public institutes and coordinating this effort at a technocratic level.

To achieve the purposes of the law, the Water Corporation, in cooperation with specialized and concerned bodies, would exercise the following function:[10]

1. Draw the general policy for environmental protection, with the strategy and plans for implementation;
2. Monitor environmental parameters through the laboratories designated by the Board;
3. Prepare specifications and parameters for environmental components;
4. Carry out research and studies relevant to the environment;
5. Monitor utilities, public and private activities including projects and companies to ensure their compliance with environmental parameters and specifications;
6. Lay down regulations, specifications, and environmental conditions;
7. Supervise and conduct environmental impact assessment for projects;
8. Lay down rules for handling hazardous waste;
9. Lay down rules for establishing Nature Reserves;
10. Prepare plans for environmental emergencies.

The Water Corporation is an autonomous body, a legal entity with financial independence, and is the authorized body responsible for environmental protection within all sectors of the developmental scheme. Public and private entities must comply with its regulations and decisions. Courts deal with infringements of the law, but the Director of the Corporation can close any premises if the infringement is severe until the matter is dealt with by the courts or the infringement is rectified.

A Higher Council for Environment Protection was established to be headed by the Minister of Municipal and Rural Affairs and Environment, consisting of 21 members from the public sector mostly and the private sector. The council, among other things, has the following mandates:

1. Approve the national environmental policy and strategy;
2. Approve environmental specification and standards;
3. Propose environmental legislation;
4. Issue regulations and decisions to add to the Environment Protection Law;
5. Approve plans to deal with environmental disasters.

Other public organizations are also involved in environmental planning and implementation of environmental protection measures, each within its own mandate.

The Ministry of Planning (MOP) is the official government body entrusted with and responsible for channeling funding from donor countries and organizations to Jordanian agencies and organizations.

However, an Environment Unit (EU) in the MOP was placed in the Water, Environment and Tourism Directorate in December 1994 as a result of growing demand from governmental and non-governmental organizations that are applying for or implementing environmental projects. The EU/MOP is primarily responsible for reviewing and commenting in coordination with related institutions, on the increasing number of environmentally related projects and proposals that are submitted to the MOP and financed by foreign donors. The role of the EU/MOP is often described as that of a facilitator between the needs of technical institutions and the requirement of donors.

Jordan's water situation is so severe that demand management (also called use reduction) is necessary to save water.[11] The greatest potential for saving water through reduced use involves three options: municipal water conservation; expanded use of water-saving equipment, especially on farms; and changed cropping patterns. Most of the savings can be accomplished with small public expenditures, although they may have strong political and economic consequences. The costs of implementing the options, such as adoption of drip irrigation equipment, are borne mainly by the private sector.

Policies Affecting Water Supply

These policies have concentrated on the following options:[12]

1. Maintaining environmental protection of water;
2. Building reservoirs;
3. Enhancing water distribution networks and other related infrastructure;
4. Treating and reusing wastewater;
5. Preventing leaks in the municipal delivery systems;
6. Controlling the use of groundwater.

The experience of Jordan in applying these options does not seem to have been effective in solving the shortage of water, although these options were not fully implemented. In fact, various causes have prevented full compliance, including lack of financial and human resources and higher policy priorities. As a result, the problem of a water deficit persists and deepens.

Areas of Priority Actions

No single area of action holds a solution for Jordan's grim dilemma of water shortage. The answer must be sought by developing a comprehensive water management plan in five priority action areas listed below.[13]

1. Strengthen the capability of the Ministry of Water and Irrigation so it can develop and fully implement sound water policies, programs, and services to the people – its customers – more effectively, efficiently, and responsively;
2. Encourage appropriate private sector participation in water resource management and pollution prevention and control;
3. Reduce the water demand by all feasible means. Make water conservation a part of everyday living, an issue of constant awareness throughout society – among the political, public, and private sector groups, and the adults and children in Jordan's homes;
4. Create real incentives to encourage efficient water conservation and discourage waste, including enforcing fully the existing regulations on water use, and developing legislation to close gaps in the law;
5. Build and maintain a public opinion setting in which knowledge of this vital resource and the means of conserving it stay on the agenda of groups and individuals throughout the land.

Jordan's water resources have always been scarce, but they have been stretched to meet the needs of the country by extracting groundwater beyond amounts replaced and by rationing municipal and irrigation supplies. However, the demand is influenced by increased irrigation, rapidly increasing population, and industrial development.

Increasing water availability requires searching for alternatives to technical solutions. These alternatives involve managing water supply and demand such as:

1. Capture or harvest storm and surface water flows.
2. Reuse domestic and industrial waters and wastewaters.
3. Conserve water by repairing leaks, suppressing evaporation, managing watersheds, and other means.
4. Develop new resources, such as brackish waters.
5. Protect the quality of existing supplies to preserve their use.

Water resource managers usually favor supply approaches, which are simpler to implement. These approaches deal more with physical facilities than with people.

The major supply options for Jordan include developing dams for storing water, improving conveyance and distribution structures, desalinizing brackish water, harvesting upland waters, storing groundwater, reducing leaks, and reusing wastewater. Nationwide, most of the initial increased supply would come from reusing wastewater, building new dams, reducing losses from leaks, and desalination of brackish waters. It is true that these sources will be essential in the long run.

Groundwater basins are the principal victims of short-term policies. They are over-exploited and in several cases unlicensed. Jordan is one of the countries that suffer most from such a situation.

At the economic level, further stress is exerted on water policies. Most of the economies in the region are in a transitional state between public sector dominance and ongoing privatization. The long history of water subsidies, which provided a base for family economies, is difficult to change. An innovative economic tool should be developed to safeguard social security while implementing strategies aiming at providing a concrete base of water security for future generations by redefining the value of environmental resources.

Concluding Remarks

Water resources management in Jordan is subject to justified criticism. The planning is unsustainable, short-termed, and unfair. There is a general lack of adequate national water policies. Most of the region's countries have developed national policies for water management, but these cycles were not completed or implemented in a manner that is environmentally sound and ecologically sustainable. In addition, these activities were not consolidated or coordinated with other water-based functions.

The distribution of water among households, agriculture, and industry is questionable. Agriculture consumes around two-thirds of supplies. Some of the crops are low-income high-water-consuming crops that contribute to the unsustainable use of a scarce resource. In Jordan, agriculture contributes less than 5% to the GDP[14] but consumes around 70% of water resources.

There are social causes also for the emergence of water security problems. These include the ever-increasing population growth which puts further stress on the already scarce water supplies. Provided the current rate of growth continues, a wide and dangerous gap between supply and demand will emerge and reach a point that makes its bridging a formidable and technically unfeasible task.

This chapter concludes that priorities for making the most of Jordan's water resources should be given to options affecting water demand on two frontiers. The first is adopting a pricing strategy including pricing reforms, subsidy modifications, and water conservation subsidies. The second concerns managerial options, including improving metering, billing and revenue collection, redistribution among consumers, educational initiatives, and human resources development for employees dealing with water issues. The main theme of this essay is that in the absence of options for increasing water supplies, greater efficiency in managing available water resources – including water consumption, pricing, and integrating economic and social reforms – is the only option left. An economic approach for making the most of Jordan's water resources will have a greater impact on enhancing the water status in Jordan.

It also concludes that technical options, which are usually directed to improve water supply, are infeasible in the short run. The message is clear: An economic approach, which is about optimum use and management of scarce resources, is more efficient than technical options that aim at increasing water supply. While water technologies are advancing, the lack of human and financial resources will still make Jordan lag behind.

Since Jordan is a small developing country that shares borders with countries that affect part of its water supply, the search for efficiency of water use and management should consider interregional microeconomics as a research priority. The adequacy of economic theory and the availability of data may better describe regional cooperation in water consumption and management than desalination of water, cloud seeding, or other technical options.

Moving towards more efficient and sustainable water use requires major changes in the way water is valued. Setting prices closer to the real cost of supplying water is a key component of both urban and industrial conservation.

Research on water use and management is vital in conserving and reallocating water consumption so that per capita consumption of water will increase. The problems of cost of water, geographical aggregation, assigning relative weights to social, economic and environmental concerns of water resources are of prime importance.

Notes

1 E. D. Buskirk et al., A Water Management Study of Jordan, Unpublished Technical Report #4 (Amman, 1992); A. Garber and E. Salameh (eds.), *Jordan's Water Resources and their Future Potential* (Amman: Friedrich Ebert Stuftung, 1992); A. Z. Ghezawi and H. S. Dajani, *Jordan's Water Sector: Facts Manual* (Amman: Royal Scientific Society, 1995); E. Salameh, and H. Bannayan, *Water Resources of Jordan: Present Status and Future Potentials* (Amman: Friedrich Ebert Stuftung, 1993); S. C. Lonergan and D. B. Brooks, *Watershed: The Role of Fresh Water in the Israeli–Palestinian Conflict* (Ottawa: International Research and Development Centre, 1994).

2 J. A. Allan and J. H. Court (eds.), *Water, Peace and the Middle East: Negotiating Resources in the Jordan Basin* (London: St. Martin's Press, 1996); R. Alwyn and R. Rouyer, *Turning Water into Politics: The Water Issue in the Palestinian–Israeli Conflict* (London: St. Martin's Press, 1999); P. Beaumont, The Myth of Water Wars and the Future of Irrigated Agriculture in the Middle East, *Water Resources Development*, Vol. 10, No. 1, pp. 9–22, 1994; F. Hof, The Yarmouk and Jordan Rivers in the Israel–Jordan Peace Treaty, *Middle East Policy,* Vol. 3, No. 4, pp. 47–56, 1995; S. Libiszewski, *Water Disputes in the Jordan Basin Region and their Role in the Resolution of the Arab–Israeli Conflict* (Zurich: Center for Security Studies and Conflict Research, Occasional Paper No. 13, 1995); M. R. Lowi, *Water and Power: The Politics of a Scarce Resource in the Jordan River Basin* (Cambridge: Cambridge University Press, 1995); and D. Wishart, An Economic Approach to Understanding Jordan Valley Water Disputes, *Middle East Review,* Vol. 4, No. 21, pp. 45–53, 1989.

3 Jordan, Central Bank of Jordan, Annual Report 2002 (Amman: Central Bank of Jordan, 2002), p. 8.

4 Ibid., p. 12.

5 *Jordan Times*, January 29, 2001.

6 Jordan, Ministry of Water and Irrigation, Annual Report 2002 (Amman: Ministry of Water and Irrigation, 2002), p. 17.

7 *Jordan Times*, January 29, 2001.

8 A. Z. Ghezawi and H. S. Dajani, Jordan's Water Sector: Facts Manual (Amman: Royal Scientific Society, 1995), p. 11.

9 *Jordan Times*, August 15, 2000.

10 Ibid.

11 E. D. Buskirk., A Water Management Study of Jordan (Amman: Unpublished Technical Report #4, 1992), p. 7.

12 Ibid., pp. 1–2.

13   Ibid., pp. 3–5.

14   Jordan, Central Bank of Jordan, *Monthly Statistical Bulletin*, Vol. 39, No. 8
     (Amman: Central Bank of Jordan, 2003), p. 83.

CHAPTER 14

Water: Casus Belli or Source of Cooperation?

FRANKLIN M. FISHER

So important is water that there are repeated predictions of water as a *casus belli* all over the globe. For example, the UN-sponsored Third World Water Forum stated in August, 2001 that water could cause as much conflict in this century as oil did in the last. Said Crown-Prince Willem-Alexander of The Netherlands: "Water could become the new oil as a major source of conflict."[1] Similarly, former United States Senator Paul Simon wrote:[2]

> Nations go to war over oil, but there are substitutes for oil. How much more intractable might wars be that are fought over water, an ever scarcer commodity for which there is no substitute?

He went on to say:

> Last year American intelligence agencies told President Bill Clinton, in a worldwide security forecast, that in 15 years there will be a shortage of water so severe that if steps are not taken soon for conservation and cooperation, there will be regional wars over it.

And these are but two very recent examples of many.[3]

Such forecasts of conflict, however, stem from a narrow way of thinking about water. Water is usually considered in terms of quantities only. Demands for water are projected, supplies estimated, and a balance struck. Where that balance shows a shortage, alarms are sounded and engineering or political solutions to secure additional sources are sought. Disputes over water are also generally thought of in this way. Two or more parties with claims to the same water sources are seen as playing a zero-sum game. The water that one party gets is simply not available to the others, so that one party's gain is seen as the other parties' loss. Water appears to have no substitute, so that it can only be traded for other water.

But there is another way of thinking about water problems and water disputes, a way that can lead to dispute resolution and optimal water management. That way involves thinking about the economics of water and shows, in fact, that water can be traded off for other things. Further, it shows that cooperation in water is a far more sensible policy than is autarky (self-sufficiency in water) – provided, of course, that there is someone with whom to cooperate.

Is Water Worth War?

The late Gideon Fishelson of Tel Aviv University once remarked that "Water is a scarce resource. Scarce resources have value." He went on to point out that the possibility of desalinating seawater (together with the costs of conveyance from the seacoast) must put an upper bound on the value of water in dispute to any country that has a seacoast. This implies, for example, that the value of the water in dispute between Israelis and Palestinians lies at most in the range of a few hundred million dollars per year and is most probably far less than that. Such amounts ought not to be a bar to agreement between nations.

Fishelson's remarks were a principal impetus to the creation of the Water Economics Project (WEP). That project, begun in 1993, is a joint endeavor of Israeli, Jordanian, Palestinian, Dutch, and American scholars. Since 1996, it has been supported by the government of the Netherlands and has operated with the consent, if not the commitment, of the Israeli, Jordanian, and Palestinian governments.

The WEP finds that the value of the water in dispute is in fact far less than the Fishelson upper bound. (This corresponds to the finding that, except in years of extreme drought, desalination on the Mediterranean coast is not and will not be efficient.)

For example: In 2010, the loss of an amount of water roughly equivalent to the entire flow of the Banias springs on the Golan Heights (125 Million Cubic Meters [Mm^3]) annually) would be worth no more than $5 million per year to Israel in a year of normal water supply and less than $40 million per year in the event of a reduction of thirty percent in naturally occurring water sources. At worst, water can be replaced through desalination, so that such water (which has its own costs) can never be worth more than about $75 million per year. These results take into account Israeli policies towards agriculture.

A similar result applies to the Hasbani, which was seen as a possible *casus belli* when Lebanon proposed the taking of considerable water from it some years ago.

Note that it is *not* suggested that giving up so large an amount of water is an appropriate negotiating outcome, but water is not an issue that should hold up a peace agreement. These are trivial sums compared to the Israeli GDP or to the cost of fighter planes.

The WAS Tool: Domestic Results

The owner of water that consumes the water itself does not obtain that water for nothing. Instead, the owner incurs an opportunity cost – giving up the money that the water would bring if sold. Indeed, the questions of who *owns* water and who *uses* water can be shown to be entirely separate. Water ownership is a property right entitling the owner to the money that the water represents. The WEP built a decision support tool, a computer model called WAS (Water Allocation System). WAS divides the area to be studied into districts. Within each district, data are provided on water sources and costs of extraction, and demand curves are specified for three user classes: households, industry, and agriculture. Retreatment of wastewater is permitted, and infrastructure is specified including conveyance lines, wastewater retreatment plants and desalination facilities – such infrastructure being either actual or potential. WAS maximizes the net benefits from water (defined as the amount water users are willing to pay for water less the costs of providing the water). This powerful tool permits the user to investigate optimal water management given the values and restrictions that the user imposes. For example, the user can specify that, as in Israel, agriculture is to receive water at subsidized prices.

The WAS tool can be used domestically to advise and inform water policy or the analysis of costs and benefits from proposed infrastructure projects or of the value of new water sources. Such uses take into account the system-wide effects of decisions or projects.[4]

Here are some examples of the results of such use:

- Israel sometimes has a short-term water problem, due to drought. It does not have a long-term "crisis" regarding water quantity. Rather the long-term problem is a monetary one. At the worst, that problem can be solved by desalination which can produce unlimited water quantity but at a price too high for unsubsidized agriculture.
- But desalination (even at 60 cents per cubic meter) will not be necessary or efficient for some time to come, except in extreme drought years.
- Jordan's water problems will not be solved by obtaining a greater share of the Jordan or Yarmouk rivers without the building of substantial new conveyance infrastructure to take such water to Amman. The benefits from repair of Amman's distribution system would far exceed the required expense.

The WAS Tool: International Issues

The uses of WAS are not restricted to domestic management, however. The WAS tool can also be used in the resolution of water disputes.

• Water and water disputes can be monetized and analyzed in terms of economics, taking full account of the social or national value of water that may exceed private value. This can assist by showing the true size of the water problem, which will often not be very large.
• Further, each party can use its own version of WAS to evaluate the consequences to it of different water agreements.
• Finally, the parties can cooperate by agreeing to trade short-term permits to use water at prices that reflect the scarcity value of the water. WAS generates such prices and shows the gains from cooperation.
• Perhaps most important of all, the WAS tool can be used to guide cooperation in water and to estimate what such cooperation would be worth.

Basically, WAS-guided cooperation consists of neighbors trading "water permits" – short term access to each other's water – and doing so at efficiency prices generated by the WAS tool. Such prices reflect the values put on water by each participating entity. Since trade in water permits is voluntary, both the buyer and the seller of water permits gain from such transactions. The buyer receives water that it values more highly than the money given up to buy it; the seller receives money that it values more highly than the water it gives up in the sale. The result is a "win–win" situation.

We have estimated the gains to Israel and the Palestinians from such cooperation, and find them to exceed the value of changes in water ownership that reflect reasonable differences in negotiating positions.[5]

Figures 14.1–6 illustrate such findings and more besides. In those figures, we have arbitrarily varied the fraction of Mountain Aquifer water owned by each of the parties from 80% to 20%.[6]

The two line graphs in figure 14.1(a) show the gains from cooperation in 2010 for Israel and Palestine, respectively, as functions of ownership allocations.[7] Israeli price policies for water are assumed to be the same as in 1995, with large subsidies for agriculture and much higher prices for households and industry.

Starting at the left, we find that Palestine benefits from cooperation by about $170 million per year when it owns only 20% of the Mountain Aquifer. In the same situation, Israel benefits by about $12 million per year. As Palestinian ownership increases (and Israeli ownership correspondingly decreases), the gains from cooperation fall at first and then rise. At the other extreme (80% Palestinian ownership), Palestine gains about $84

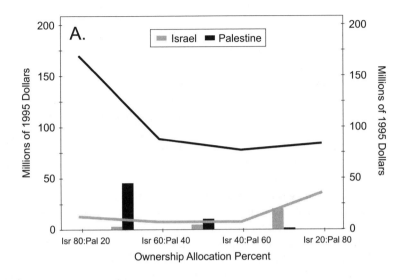

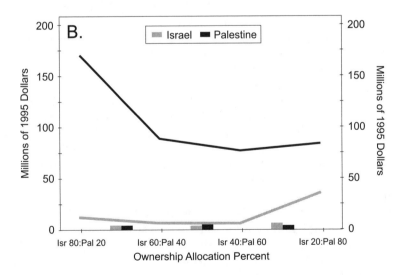

Figure 14.1 Value of Israel–Palestine Cooperation and Value of Ownership of Mountain Aquifer without (a) and with (b) cooperation (2010 – Israel Fixed Priced Policies in Effect).

million per year from cooperation, and Israel gains about \$36 million per year. In the middle of the figure, total joint gains are about \$84–\$95 million per year.

It is important to emphasize what these figures mean. As opposed to autarky, each party benefits as a buyer by acquiring cheaper water. Moreover, each party benefits as a seller by tens of million of dollars per year over and above any amounts required to compensate its people for increased water expenses.

Why do the gains first decrease and then increase as Palestinian ownership increases? That is because, at the extremes, large gains can be made by transferring water from the large owner to the other party. Palestine has larger benefits at the left-hand side of the diagram than in the middle because it can obtain badly needed water from cooperation; it has larger gains at the right-hand side than in the middle because, when it owns most of the Mountain Aquifer water, it can gain by selling relatively little-needed water to Israel (who gains as well). The same phenomenon holds in reverse for Israel – although there the effects are smaller, largely because Israel is assumed to own a great deal of water from the Jordan River.

One might suppose that the gains would be zero at some intermediate point, but that is not the case. The reason for this is as follows:

It is true that a detailed, non-cooperative water agreement could temporarily reduce gains to cooperation to zero. That would require that the agreement exactly match in its water-ownership allocations the optimizing water-use allocations of the optimizing cooperative solution. That is very unlikely to happen in practice (and, if it did, would only reach the optimal solution for a very short time, as explained below). In our runs, it does not happen for two reasons.

We have not attempted to allocate ownership in the Mountain Aquifer in a way so detailed as to match geographic demands. Instead, we have allocated each common pool in the aquifer by the same percentage split.

There are gains from cooperation in these runs that do not depend on the allocation of the Mountain Aquifer. It is always efficient for Gaza to be supplied from the Israeli National Carrier, and it is always efficient for treated wastewater to be exported from Gaza to the Negev for use in agriculture.

There are further results to be read from figure 14.1(a). The height of the various bars in the figure show the value to the parties without cooperation of a change in ownership of 10% of the Mountain Aquifer (about 65 Mm^3 per year or nearly half the Mountain Aquifer water now taken by Palestine). These are calculated by looking at the changes in ownership used in the results, so that, for example, the left-hand-most set of bars show the value to the parties of half the change from an Israeli 80%–Palestinian 20% to an Israeli 60%–Palestinian 40% allocation of ownership; the next set of bars examines the value of half the change from 60–40 to 40–60.

Note that the value of cooperation generally exceeds the value of such ownership changes. Note also, that a great deal of water is involved.

Further, now look at figure 14.1(b). This differs from figure 14.1(a) only in the height of the ownership-value bars. In figure 14.1(b), the height of those bars represents the value of shifts of 10% aquifer ownership in the presence of cooperation. That value is about $8 million per year. The lesson is clear:

Ownership is surely a symbolically important issue, and symbols really matter. But cooperation in water reduces the practical importance of ownership allocations – already not very high – to an issue of very minor proportions.

The same qualitative results hold when we examine figures 14.2(a) and 14.2(b). These differ from figures 14.1(a) and 14.1(b). respectively, in that Israeli Fixed-Price policies (FPP's) are assumed not to be in effect and water is sold to users at the efficiency prices generated by WAS.

There are three differences from the figures 14.1(a) and 14.1(b) that are worth discussing.

While the value to Palestine of ownership changes *without* cooperation remain the same as before (as they must, since Palestine receives exactly the same water as before), the gains to Israel are reduced, but reduced significantly only when Palestine owns the lion's share of the aquifer.

The value of ownership changes *with* cooperation is even smaller in figure 14.2(b) than in figure 14.2(a), about $3–$4 million per year.

The gains from cooperation are not much different, and the difference is interesting.

One of the issues that might arise in contemplating a cooperative agreement of the type described is as follows: If Israel subsidizes water for agriculture,[8] then the demand for water by Israeli farmers will rise. Since this increases water scarcity, it will increase the efficiency prices of water and hence the prices to Palestinian consumers. Does not that mean that there will be continual negotiation over Israel's price policy? The answer turns out to be in the negative.[9]

Figure 14.3 shows the difference to each party made by Israeli FPP's in the context of a cooperative agreement. The negative effects on Palestine are very small at worst. They are about $4 million with Israeli ownership of 80% of the aquifer, and, indeed, with increases in Palestinian ownership, the effects rise toward zero, eventually even becoming positive. This occurs because the use of FPP's also increases the price that Israel must pay to obtain Palestinian water, and, with increasing Palestinian ownership, the amount of such purchases rises.

Of course, the corresponding effect on Israel itself is in the other direction. The effect of FPP's on Israel starts off negative and becomes increasingly so. It must be remembered, however, that this is the price of having the FPP's, particularly of subsidizing agriculture. Presumably,

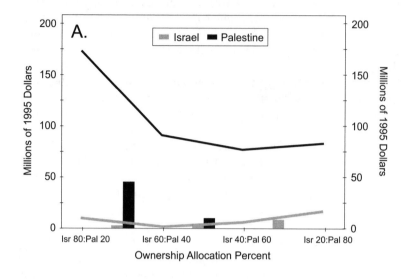

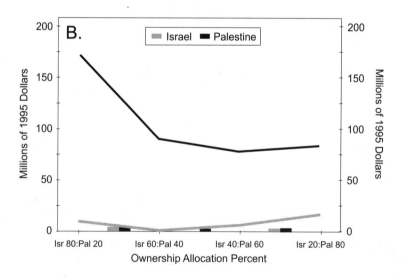

Figure 14.2 Value of Israel–Palestine Cooperation and Value of Ownership of Mountain Aquifer without (a) and with (b) cooperation (2010 – No Fixed Priced Policies).

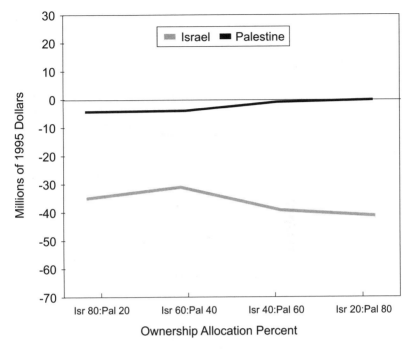

Figure 14.3   Effect of Israeli Fixed Priced Policies on Total Surplus for Israel and Palestine Under Cooperation: 2010.

Israel's policy makers would consider that there is an added social gain from doing so – a gain not reflected in the calculations shown.

Figures 14.4–6 (pages 194–96) show similar results for 2020. As we should expect, all the monetary figures are greater (for example, total gains from cooperation range from $116 to $232 million per year instead of from $84 to $184 million per year), but the qualitative conclusions are the same. Note in particular that the value of an ownership shift of 10% of the Mountain Aquifer under cooperation is only about $3.5–$11.5 million per year. While water will be more valuable in 2020 than in 2010, mostly because of population growth, the value of cooperation will still generally be higher for each party than a gain of ownership of 10% of the aquifer. (Note that, as in all such figures, the comparison must be made between a line of one color and bars of the same color.)

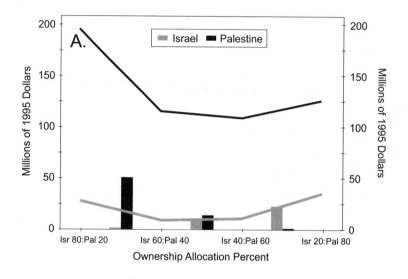

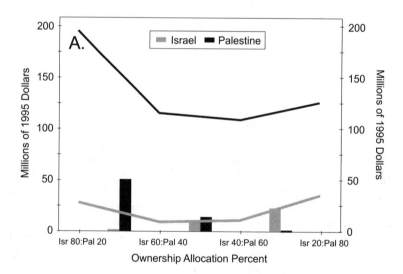

Figure 14.4 Value of Israel–Palestine Cooperation and Value of Ownership of Mountain Aquifer without (a) and with (b) cooperation (2020 – Israeli Fixed Priced Policies in Effect).

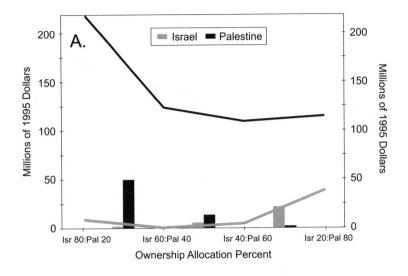

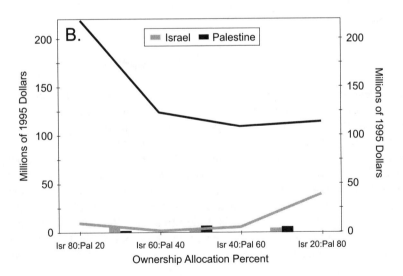

Figure 14.5 Value of Israel–Palestine Cooperation and Value of Ownership of Mountian Aquifer without (a) and with (b) cooperation (2020 – No Fixed Priced Policies).

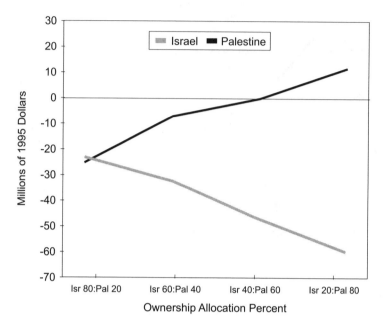

Figure 14.6   Effect of Fixed Priced Policies on Total Surplus for Israel and Palestine Under Cooperation: 2020.

The Real Benefits from Cooperation

The greatest benefits from cooperation may not be monetary. Beyond pure economics, the parties to a water agreement would have much to gain from an arrangement of trade in water permits. Water quantity allocations that appear adequate at one time may not be so at other times. As populations and economies grow and change, fixed water quantities can become woefully inappropriate and, if not properly readjusted, can produce hardship. A system of voluntary trade in water permits would be a mechanism for flexibly adjusting water allocations to the benefit of all parties and thereby for avoiding the potentially destabilizing effect of a fixed water quantity arrangement on a peace agreement. It is not optimal for any party to bind itself to an arrangement whereby it can neither buy nor sell permits to use water.

Moreover, cooperation in water can assist in bringing about cooperation elsewhere. For example, as already indicated, the WAS model strongly suggests that, even in the presence of current Israeli plans, it would be efficient to have a water treatment plant in Gaza with treated effluent sold to Israel for agricultural use in the Negev where there is no aquifer to pollute.

(Indeed, we are informed that since this suggestion arose in model results, there has been discussion of this possibility.) Both parties would gain from such an arrangement. This means that Israel has an economic interest in assisting with the construction of a Gazan treatment plant. This would be a serious act of cooperation and a confidence-building measure.

Problems and Conclusions

Naturally, there are a number of issues that arise as to such a scheme. Chief among them is that of security. What if one of the partners to such a scheme were to withdraw? Of course, such withdrawal would be contrary to the interest of the withdrawing party, but, as we have sadly seen, people and governments do not always act in their own long-run self-interest.

The main cost of such a withdrawal to the other party would occur if that party had failed to build infrastructure that would be needed without cooperation but not with it. In the case of Israel and Palestine, that risk would be chiefly Palestinian, since they, but not Israel, would need desalination plants in the absence of cooperation but not in its presence. Israel, by contrast, already has a highly developed system of water infrastructure and any decision to build desalination plants does not depend on a decision to cooperate or not cooperate with the Palestinians.

Interestingly, however, we have found that the necessity for a Palestinian desalination plant in Gaza is not primarily due to the expected increase in the Gazan population. Rather it is because, with the current allocation of Mountain Aquifer water, it will be desirable for Palestine to desalinate seawater at Gaza and pump the desalinated water *uphill* to supply the southern West Bank.[10] Of course, such necessity disappears under cooperation. In the case of an Israeli withdrawal from a water-permit trading agreement, therefore, Palestine could relieve its problem by overpumping its allotment of Mountain Aquifer during the time it takes to construct the needed desalination and conveyance facilities.

For Israel, at least, therefore, cooperation is clearly a superior policy to autarky. In an atmosphere of trust, cooperation would be likely to benefit Palestine even more. But, of course, such an atmosphere does not now exist. Cooperation requires a partner, and, at present, that does not appear to be immediately likely. Despite this, I continue to believe that cooperation is both valuable and possible. As already discussed, water is not worth conflict and can become an area for confidence-building measures. Further, if autarky is truly desired, then one should simply build desalination plants as needed. Autarky in naturally-occurring water is a foolish policy except as a money-saving device – and the money it saves is not great.

Every country with a seacoast can have as much water as it wants if it chooses to spend the money to do so. Hence, every country with a seacoast can be self-sufficient if it is willing to incur the costs of acquiring the necessary water. Further, cooperation in water, is more efficient than desalination (at least for Israel and Palestine), and produces a win–win situation, benefiting all the parties involved. Disputes over water are merely disputes over costs, not over life and death. They are not worth war and can be resolved to the benefit of all parties.

Acknowledgments

This chapter draws on the work of a large number of people involved in the Middle East Water Project discussed below – too many to thank by name – particularly the co-authors of F. M. Fisher, A. Huber-Lee, I. Amir, S. Arlosoroff, Z. Eckstein, M. Haddadin, S. G. Hamati, A. Jarrar, A. Jayyousi, U. Shamir, and H. Wesseling (2005) *Liquid Assets: An Economic Approach for Water Management and Conflict Resolution in the Middle East and Beyond* (Washington: RFF Press, 2005), which gives a far more complete and detailed discussion than is possible here. I am greatly indebted to the government of the Netherlands for its support of the project. The views expressed are not necessarily those of any government or person other than myself. I am grateful to Brian Palmer and, especially, to Annette Huber-Lee for assistance but retain responsibility for error.

Notes

1   Reuters interview reported on Environmental News Network, August 13, 2001.

2   P. Simon, "In an Empty Cup, a Threat to Peace," *New York Times*, August 14, 2001.

3   For example, see pp. 56–7, 59–60 in M. T. Klare, "The New Geography of Conflict," *Foreign Affairs* 80 (2001): 49–61.

4   For a far more detailed description of WAS and examples of its use, see Fisher et al. (2005). See also F. M. Fisher, 2001, "The Economics of Water and the Resolution of Water Disputes." In *Negotiation Over Water: Proceedings of the Haifa Workshop, 1997* (U. Shamir, ed.), International Hydrological Programme, IHP-V, Technical Documents in Hydrology, No. 53, UNESCO, pp. 71–99, Paris and F. M. Fisher, S. Arlosoroff, Z. Eckstein, M. Haddadin, S. G. Hamati, A. Huber-Lee, A. Jarrar, A. Jayyousi, U. Shamir, and H. Wesseling (2002), "Optimal Water Management and Conflict Resolution: The Middle East Water Project." *Water Resources Research*, vol. 38, no. 11 (November, 2002).

5   There are qualitatively similar results for cooperation between Israel and Jordan and results for cooperation among all three parties, but space does not permit analyzing them here. See Chapter 8 of Fisher et al. (2005).

6   Mountain Aquifer water is water under the hills of the West Bank and of Israel. We have equally arbitrarily assumed in these figures that Israel owns 100% of the water of the Jordan River. None of these assumptions is intended to convey a political message as to the appropriate allocation of water ownership.

7   Here and later, the results refer to a year of normal hydrology. Results for drought years are not qualitatively different, although all numbers are larger.

8   For the sake of exposition, I examine Israeli FPP's, but a similar issue would
    arise if Palestine were to subsidize water for agriculture.
9   There is, of course, a different and possibly more important issue. If Israeli
    farmers receive water at subsidized prices, their costs will be less, thus enabling
    them to compete more effectively in the markets for agricultural outputs. But
    that would be true of any subsidy to agriculture, not just a water subsidy. It
    has nothing to do directly with water price policy as such.
10 See Fisher et al. (2005).

CHAPTER 15

Water, Demography, and Future Economic Development in the Triangle: Jordan, Israel, and the Palestinian Territories

ONN WINCKLER

Following the struggle over land, the issue of water constitutes the most crucial dispute in the Arab–Israeli conflict, particularly between Israel and the Palestinians. Alwyn Rouyer opened his book by stating that "a peace treaty will not be reached between Israel and the Palestinian Authority without an agreement that provides an equitable distribution of the water resources . . . which these two peoples share."[1] Moreover, the dispute about water always was, and still is, a major factor in the relationships between other Middle Eastern countries, such as between Turkey, Syria, and Iraq on the Euphrates water; or between Egypt, Sudan, and Ethiopia on the Nile water. However, despite the massive discussion on the issue of the severe water shortage in the Triangle, namely, Jordan, Israel and the Palestinian territories, it should be emphasized that more than half of the water use in the Triangle is not for domestic and industrial use, but rather for agricultural purposes.[2]

Attitudes towards the Middle Eastern Water Shortage

Overall, it can be concluded that during the past three decades, four prominent attitudes towards the Middle Eastern water dispute have developed among academic researchers, as well as among politicians and decision makers. The first and perhaps the most prevailing attitude is that the increasing scarcity of water has constituted, and will continue to constitute, a prominent factor in the Arab–Israeli conflict, particularly between Israel and the Palestinians. For example, in *The Politics of Water in the Middle East*, Martin Sherman wrote as follows:

> The conflict between Arab and Jew in the Middle East has been one of the most acrimonious and protracted . . . Deep-rooted, divergent ideological convictions, strong . . . religious beliefs in mutually exclusive God-given rights, and seemingly irreconcilable national aspirations all combine to make for a dispute that abounds with innumerable issues of seemingly intractable contention. Few, however, have greater potential for deadlock than the question of water.[3]

The opposing attitude to the Middle Eastern water shortage can be summarized as "Hydro-Paranoia." The main argument of this position is that the scarcity of water in the Middle East has resulted first and foremost from improper use of water resources, including the use of inefficient irrigation methods and waste of large quantities of water for domestic use in both urban and rural areas. Berthold Hoffmann claimed that "efforts need to focus on decreasing demand for water, or, at least, make more efficient use of already available resources."[4] A key to this strategy, claims Hoffmann, is "water pricing."[5] Obviously, a certain quantity of water for drinking and other domestic use as well as for health needs should be supplied to the population "at a cost affordable by all." However, beyond that level, water should be sold at its "true price," including all the expenses entailed in production, channeling, and taxation. In addition, the adoption of modern cultivation and irrigation methods, combined with measures to increase efficiency in industry and domestic use will eventually bring about a solution, at least partially, to the problem of water scarcity in the region. Moreover, it should be noted that much of the water demand, particularly in the urban centers, is due to low prices, particularly in Israel. Thus, raising the water prices to "the true price," as Hoffman suggested, will naturally bring about a considerable decrease in demand, particularly for domestic use.

The main argument of the third attitude is that the water issue is not a "zero-sum game" and that through cooperation in various water projects between the Triangle, as well as between other Middle Eastern countries, in such areas as desalination and construction of new dams, the problem of water shortage can be solved, or at least eased to a large extent. Thus, many seminars, international conferences, and bilateral meetings have been held following the 1991 Madrid Conference in order to enhance cooperation within the Triangle on the issue of water, perhaps more than for any other economic issue. Good examples of this approach are the various Israeli–Jordanian water programs which unfortunately until now were not implemented. This cooperation, in parallel with the adoption of more effective and modern methods in both the agricultural sector and in domestic use, will serve to ease the water shortage to a large extent, thereby reducing, or even eliminating, its potential as a "time bomb."

The fourth approach is that water should be viewed as a commodity like any other commodity and not as a crucial national resource. According to this approach, the Israeli policy towards the dispute over water with its neighboring countries should be "measured in economic terms, as they would essentially amount to the difference between the cost of supplying the water foregone and the cost of water produced to substitute for them. This cost differential is relatively small, when compared to the cost of conflicts or the Israeli GDP," as claimed by Eran Feitelson.[6]

An Alternative Approach: The Economic Solution

Here, I present an alternative approach. My basic argument is that the Triangle governments, similar to those of the other Middle Eastern non-oil economies, with the exception of Turkey, must realize that water scarcity for large-scale agriculture will eventually be an inescapable reality in this region. Such would be the case even if all existing and future projects for increasing water potential through desalination or imported water, either from Turkey or from Egypt, were to be implemented. This is due simply to the rapid population growth and the accompanying wide-based age pyramid that can be expected to continue for at least the coming four decades, even if fertility rates sharply decline.

According to the recent medium-variant projections of the UN Department of Economic and Social Affairs, by the year 2025, Jordan's population will amount to 8.7 million, while that of Israel will reach 8.5 million.[7] According to the US Census Bureau, the Palestinian population in the West Bank and Gaza Strip will amount to 5.8 million by the year 2020.[8] Thus, within one generation the Triangle population will reach more than 23 million, almost twice its current size.

Moreover, due to the expected increase of water for domestic use as a result of improving living standard, and in line with the continuing rapid population growth, the water scarcity for large-scale agriculture should no longer be treated as an isolated issue, but rather as part of a much broader issue – the performances of the Triangle economies and their available financial and energy resources. As such, the Triangle governments should concentrate their efforts and investments on determining the optimal means to improve their current economic performances either through agriculture or through other economic sectors.

In line with this basic attitude, the main question that the Triangle leaders need to address is whether agriculture constitutes the optimal investment for enhancing their economic performances, particularly through the use of expensive desalinated water. It seems that the answer to this question is no. Moreover, it should be taken into account that a substantial increase in the Triangle's water resources through massive desalination requires extensive energy resources, either oil or gas, which, in contrast to the Arab Gulf oil states, are not available in large enough quantities in this area. In this regard, I would like to quote Professor Tony Allan (School of Oriental and African Studies, London) who said in early 2001 that: "Having a strong and diverse economy is more important than being short of water . . . you can remedy a shortage of water through a diversified economy."[9]

During the past decade, the two major economic problems of both Jordan and the Palestinians, as well as Israel, albeit to a lesser extent, are the huge balance of trade deficits and the severe scarcity of new work opportunities available to the rapidly growing workforces. In 2000, the balance of trade deficit in Jordan amounted to $2.64 billion, representing one-third of Jordan's total GDP ($8.34 billion).[10] A report of the economic weekly *MEED* from mid-2001 noted in this regard that: "Overall, Jordan's trade record has been poor, with a sharp rise in imports and a disappointingly slow growth in exports leading to a 22% rise in the deficit."[11] Among the Palestinians, the percentage of the balance of trade deficit of the total GDP was even higher than in Jordan due to the steady deterioration of the Palestinian economy, particularly since the onset of *al-Aqsa Intifada* (September 2000).

According to the latest available data, the unemployment rate in Israel was more than 10%. Among the Jordanians, in 2001, according to official figures, the unemployment rate was 15.7%.[12] However, according to unofficial sources, the rate was actually much higher, reaching over 20%.[13] Among the Palestinians, the unemployment rate was already high before the outbreak of the *al-Aqsa Intifada*, estimated at 12.7% in 1999, with 10.1% in the West Bank and 18.8% in the Gaza Strip. Since then, as is well known, due to the deterioration in the political–security situation, the level of unemployment has continued to rise. According to the UNSCO (United Nations Special Coordinator in the Occupied Territories) report from Spring 2001, "within days of the onset of the crisis, the core unemployment rate rose . . . to nearly 30% of the labor force."[14]

Tourism as an Alternative Option for Rapid Growth in the Triangle

The basic thesis of this essay is that in order to achieve high rates of economic growth in the long term, the Triangle, particularly Jordan and the Palestinians, must adhere to the basic rule of the modern economy – the comparative advantage. This means that they should concentrate their

efforts in fields where they have clear advantages over other worldwide
economies, the most prominent and profitable of which is the tourism
industry. As Gad Gilbar and Winckler illustrated in the case of Jordan,[15]
and Joel Mansfeld and Winckler regarding Egypt,[16] the Middle East,
including the Triangle, offers some very important and unique advantages
in the area of tourism over other tourism destinations worldwide. First and
foremost is religious tourism. Second, the Middle East is known as "the
cradle of civilization," being one of the richest regions worldwide in
cultural, historical, and heritage sites. Third, the Middle East region is
simultaneously a place for winter in the north and summer in the south,
providing opportunities for a variety of climate destinations in one trip.
Fourth, the Middle East has great potential in the area of sun, sand, and
sea tourism throughout all the seasons of the year.

For the Arab non-oil economies, particularly Jordan and the
Palestinians, the tourism industry has many advantages, of which the
following are the most prominent. First, governmental revenues can be
expanded through taxation. Second, the tourism industry serves as a major
catalyst for the development of other economic sectors, mainly construction, transportation, textile, agriculture, and fisheries, insofar as they
supply the goods and services used in the tourism industry. Third, the
tourism industry, in particular, promotes foreign investments, especially in
the construction of new hotels and tourism sites. Fourth, the tourism
industry seems to be more effective than other industries in generating a
considerable number of new jobs in a wide variety of skill levels.

Conclusion

This is not all meant to say that Triangle policy makers should not devote
any resources to water projects and should instead direct the vast majority
of their investments to the tourism sector. However, in addition to channeling resources to saving water used for both agricultural and domestic
use, in parallel to increasing water prices, the major effort and investment
should be directed to the most beneficial economic sector in this region –
the tourism industry and its related services and industries. This is because
the tourism industry would bring about the highest economic return for
their investments in both foreign exchange earnings and the creation of
additional new work opportunities.

It should be emphasized, however, that the tourism industry in the
Triangle should not be viewed as a "zero sum game," meaning that an
increase in the tourism activity of one country is not necessarily accomplished at the expense of the other two. In fact, many of the tourists to this
area intend to visit all three entities in the same trip. However, the other
side of this coin is the fact that in comparison to other industry and agri-

culture, tourism is very sensitive to any political or security upheaval since it can not be "exported." Moreover, the security situation, both internal and external, in one country affects tourism activities in the entire region, as well illustrated in the entire Middle East region, but particularly in the Triangle. Thus, tourism cooperation within the Triangle has the potential to boost the number of tourists in the region for the benefit of all three entities. But, the first and the most important pre-condition is political stability and a sense of security in the Triangle. Thus, the inescapable conclusion is that a stable peace as well as economic cooperation within the Triangle are no longer a matter of option, but simply a matter of economic survival, even for the short-run.

Notes

1 Alwyn Rouyer, *Turning Water into Politics: The Water Issue in the Palestinian–Israeli Conflict* (London and New York, 2000), p. 1.

2 *The World Bank, World Development Report–1999/2000* (published for the World Bank by Oxford University Press, 2000), p. 246, table 9.

3 Martin Sherman, *The Politics of Water in the Middle East: An Israeli Perspective on the Hydro-Political Aspects of the Conflict* (London and New York, 1999), p. 1.

4 Berthold Hoffmann, Hydro Paranoia and its Myths: The Issue of Water in the Middle East, *Orient*, Vol. 39, No. 2 (1998), p. 261.

5 Ibid.

6 Eran Feitelson, The Ebb and Flow of Arab–Israeli Water Conflicts: Are Past Confrontations Likely to Resurface? *Water Policy*, Vol. 2 (2000), p. 358.

7 UN Department of Economic and Social Affairs, *World Population Prospects: The 2000 Revision* (New York, 2001), <http://www.un.org/esa/population/wpp2000at.pdf>.

8 The U.S. Department of Commerce, U.S. Census Bureau, <http:/www.census.gov>.

9 *MEED* (*Middle East Economic Digest*), February 9, 2001, p. 31.

10 *MEED*, April 20, 2001, p. 41.

11 *MEED*, June 8, 2001, p. 25.

12 The Hashemite Kingdom of Jordan, Department of Statistics, Employment and Unemployment Survey, February 2001, Principal Report (Amman: April 2001), p. 23.

13 *Jordan Times* (Amman), November 18, 1999.

14 UNSCO, The UNSCO Report on the Palestinian Economy, Spring 2001.

15 Gad G. Gilbar and Onn Winckler, The Development of the Jordanian Tourism Industry, in Joseph Ginat and Onn Winckler (eds.), *The Jordanian–Palestinian–Israeli Triangle: Smoothing the Path to Peace* (Brighton & Portland, 1998), pp. 178–95.

16 Yoel Mansfeld and Onn Winckler, Options for Viable Economic Development through Tourism Among the Non-Oil Arab Countries: The Egyptian Case, *Tourism Economics*, Vol. 10, No. 4 (2004), pp. 365–388.

Part IV

A Role for Agriculture

CHAPTER 16

High Income Innovative Crops and Optimal Fertigation System: The Solution for High Farm Income under Water Shortage in the Jordan Valley

ZVI KARCHI

The Jordan Valley stretches from the Sea of Galilee to the Red Sea. The area is divided into two parts: the northern, between the Sea of Galilee and the Dead Sea, and the southern, Wadi Arava, between the Dead Sea and the Red Sea.

The area is part of the Afro-Syrian Rift Series and is one of the deepest regions in the world: 200 m below sea level in the north and –395 m around the Dead Sea. The topography creates conditions of "High Energy," which is the region's most distinct agricultural advantage, allowing production of most crops from late fall through the winter to spring (October to early May) when prices, in most northern hemisphere markets, are the highest. The main contributors to this "High Energy" are: high day and night temperatures, low rainfall, low relative humidity and high solar energy. [1] Average daily (maximum) temperatures range between 21–35°C with average nightly (minimum) between 9–13°C. Average temperatures in the southern regions exceed those of the north by 2–4°C. There are only a few days with freezing temperatures in the north and almost none in the southern part, especially in close proximity to the Dead Sea. Duration of the freezing temperatures is usually short, mostly in the early morning hours.

Rainfall decreases in a southerly direction, being 250–300 mm annually in the north and 50–120 mm in the south. Relative humidity is low ranging between 35–50%. Low rainfall is associated with low cloudiness resulting in high solar radiation. This comprehensive package of climatological features contributes in many ways to agricultural production: high temperatures optimize and shorten crop life cycles resulting in high value produce; low relative humidity minimizes disease incidence and allows efficient, timely pest control; solar radiation improves fruit quality, enhances coloration, increases sweetness, and prolongs shelf life of vegetables, flowers and fruits.

Soils, especially in the southern region, are mostly light-textured and in many areas of sandy origin. These soils have deep drainages allowing development of deep root systems, ensuring the plants with an optimal supply of soil water throughout their life cycle.

Water is the life-sustaining component for plants; it is the carrier of nutrients absorbed by plants from soil water solution. Root systems develop downward following water stress gradients. Plants grow and are turgid due to a continuous water-column in the stems, branches, and leaves, allowing the performance of vital activities – transport of nutritive elements from the roots to the photosynthesizing leaves and moving photosynthates (carbohydrates produced by photosynthesis) from the leaves to the developing tissues. The plant maintains the continuity of the water column by the mechanism of evapo-transpiration – release of water vapor through specialized apertures (stomata), usually located on the lower side of the leaves, with its continual replacement from water absorbed from the soil by roots.[2] The rate of evapo-transpiration (and hence water loss) is temperature dependent. In the Jordan Valley, rates of evapo-transpiration during the cooler growing season average between 2–5 m^3/day/dunam[3] covered with vegetation, or 200–400 m^3/dunam/80 days of crop growing cycle. However, during the hot summer months (35°C and above) evapo-transpiration could range between 600–800 m^3/dunam/80 days. Under such high rates of water consumption, crop plants could not survive without irrigation. Water then, is the limiting factor for agricultural production in the Jordan Valley.

There are mainly two factors to consider in plant soil water relationships. First, water sources in the region are scarce and piping of water from the north is costly. Second, soil is a good water reservoir, with water held in part by soil particles and the excess irrigation water moving downward to the underground water.

The amount stored, i.e., the water field capacity, depends on soil texture. The magnitude of water at field capacity at heavy or medium textured soils to a depth of 1.0 m could be 200–300 m^3/dunam, of which about 50% is available to plants. Soils of sandy texture have very little water holding capacity and require continuous irrigation. To insure the use of water stored at depth, it is essential to promote development of plants with deep root systems. This could be done efficiently during the vegetative period of growth by building a regulated decreased water stress extending downward. There is a large volume of data showing that plants are capable of absorbing water from a depth of 1.0 m and more.[4] Deep, extensive root systems are the main attributes of plants, allowing them to overcome water stress throughout their life cycle especially in the Jordan Valley. Plants are also capable of regulating evapo-transpiration by opening and closing stomata in response to temperature changes. The activation, by plants, of built-in mechanisms to overcome water stress is taken into consideration when designing irrigation practices.

Traditionally, cultural practices are a series of techniques designed to help the farmer to cope with soil and environmental influences on plant performance. These cultural practices include a whole array of interrelated measures, e.g., crop rotation, soil fertility enhancement, tillage, seedbed preparation, choice of cultivars within each crop, plant population size, irrigation, fertilization, and disease and insect control. The most important practice, especially under Jordan Valley conditions, is the practice of fertigation (irrigation combined with balanced fertilization). Fertigation, mostly applied by the drip system of irrigation, is the most powerful tool to design plant architecture and its potential yield and its quality.[5] The principles of fertigation apply to most crop species, but this presentation is related to annual crops. The aim of fertigation under Jordan Valley conditions is to obtain optimal yields through efficient use of the least amount of water combined with balanced optimal nutrition.

Several steps are involved in the process:

1. Soil is tilled to a depth of 40–50 cm and wet to about 70 cm, prior to planting, to allow development of extensive, deep root systems.
2. Soil is covered with polyethylene sheets, with small openings to allow emergence of plants. This mulch creates a closed water system preventing evaporation of soil moisture and activating the fertile soil layer (0–30 cm), providing continuous supply of nutriments to the plants.
3. Fertigation is applied differentially to accommodate the requirements of each stage of the plant life cycle: vegetative, fruit formation, and fruit maturation.
4. The differential fertigation could be efficiently and easily accomplished by the use of a drip system of fertigation. This revolutionary equipment was developed in Israel specifically for arid regions with water shortage.[6] It is now used on millions of hectares throughout the world. Other irrigation practices such as flooding, ditch irrigation, and sprinkling are wasteful from the standpoint of water use.

In my experience, the use of the above mentioned practices could reduce water consumption by 50–75%. For example, under traditional practices, tomatoes, peppers, melons, and watermelons require 300–500 m³/dunam for yields ranging between 4–6 tons, 2–3 tons, 2–3 tons and 3–6 tons respectively. Using the advanced technologies, water requirements range between 150–250 m³/dunam and yields for field grown tomatoes, peppers, melons, and watermelons range between 8–10 tons, 4–6 tons 4–5 tons, and 6–9 tons, respectively, with income upgraded accordingly. Water saving per 1000 dunams might range between 200,000–250,000 m³ with yields and income 50–75% higher compared to traditional practices. Walk-in plastic covered tunnels and greenhouses are even more efficient in terms of economical water use and in obtaining much higher yields earlier, with excellent quality, but this is the topic for another presentation.

Prices of early produce of traditional crops in the markets are not always sufficient to cover the extra cost of advanced technologies. High value innovative crops or cultivars with added value of unique knowledge are needed to complement the cost.[7] Choice of the specific crops for practical development should be along the following criteria:

1. There should be specific adaptability to conditions of "High Energy" in terms of low water consumption, earliness of maturation, and high yields of prime quality.
2. Economic feasibility studies and market surveys are required to indicate potential volume of sales, prices, and marketing advantages. Studies and surveys should be conducted prior to the introduction of the new crop.
3. There should be sufficient volume of agrotechnical knowledge to insure immediate implementation of the new crops in farmers' fields.

There are many examples, some of which are: long shelf-life tomatoes and cherry tomatoes, colored peppers, high quality Galia melons,[8] seedless watermelons, seedless Muscat-flavored table grapes, European summer flowers in the Jordan Valley winters, and even squabs from special heavy breeds.

The combined outcome of innovative high value crops and advanced technologies results in appreciable income increase allowing farming families to gain a higher standard of living from a relatively small farm size. This process also results in a large quantity of free water contributing to additional regional development. Thus a professional approach to agricultural development can provide a realistic solution to water shortage in the Jordan Valley. The process of technology transfer requires professional organization, i.e., a Research and Development Authority.[9] The aim of the Authority is to design and operate a scheme to transfer advanced technologies and innovative crops to upgrade farm production and income. Table 16.1 summarizes the successive steps in the scheme of technology transfer in the Jordan Valley during 1983–1988.

Table 16.1   Scheme of successive steps of technology transfer in the Jordan Valley, 1983–1988

Step	Time input (years)	Operation	Responsibility
1	3–15	Research results	Scientific Institutes
2	2–3	Regional experiment station	Scientist aid by extension
3	1–2	Model plot in farmers fields	Extension aided by scientist
4	1–3	Post harvest – marketing	Marketing agency
5	–	New regional crop	Farmers

The first step is to conduct a search in the vast store of records in agricultural research institutes for completed innovative research that might be adaptable and advantageous for the "High Energy" conditions of the Jordan Valley. Since the process of research is regulated and fully recorded, information on completed research could also be obtained from scientific institutes in regions with similar climates around the world. As a rule the authority would not contribute funding to institutional research.

The second step is to test selected topics in the regional experiment station, funded by the Authority, for adaptability to climatic and edaphic conditions of the Jordan Valley. The scientist associated with the completed research is involved with extension personnel in guiding the experiments. The time span for the adaptability tests is short, 1–2 years, to obtain clear-cut results. Research proven unsuitable is discontinued and the next in priority is pursued.

The third step is to grow the new crop in "Model Plots" operated by farmers, with the guidance of extension personnel, in their fields. The cost of the "Model Plots" is guaranteed by the Authority. The size of the "Model Plot" is relatively small, 3–10 dunams, the minimum size required to yield sufficient produce for commercial shipments to export markets.

The Authority and Israel Agricultural Marketing Agency are associated in the supervision of post harvest control, produce packing and marketing shipments, constantly monitoring the economic value and market advantages of the produce. Finally, a commodity that displays upgraded economic returns is recommended by the Authority to be grown on a wide scale throughout the region.

Summary

The combined use of scientifically-based advanced technologies and innovative crops generate higher income using appreciably smaller amounts of water. To accomplish the task of technology transfer, there is a need for national policies and an atmosphere of political coordination and cooperation.

There is a long and substantial national investment in professional organizations – research institutes and extension service – capable of generating innovative data and practical knowledge. The data and know-how stored in these organizations should be the basis for a system that allows for the selection of appropriate data-based ideas, from the finished research, for practical implementation. The use of scientific, rational, and systematic research procedures insures a long-term continual flow of technical improvements and novel crops, thus maintaining and protecting the regional investments in market advantages.

The combined contribution of advanced technologies and novel crops could amount to a 50–75% increase in farm income, compared to traditional farming and a saving of about 200,000 m³ of water for each 1000 dunams. To activate the complex scientific measures on a regional basis, there is a need for a professional organization, namely a Research and Development Authority.

The consequences of the technological development contributes to the farming family in several ways:

1. The operation of a high-income farm on a reduced area helps to raise the professional skill of the farmer and raises motivation to reach greater economic success.
2. The reduced farm area, and the use of smaller amounts of water, would enable settlement of additional families in the region.
3. Specialized farming results in a better yearly distribution of income and balanced expenditures.
4. The overall operation results in free time with far-reaching personal and communal implications – better family education and advanced professional skill for the farmer. There may also be time for recreational activities.

Notes

1   A. Bitan, The climate of the Judean and Samaria lower eastern hill slopes and central Jordan Valley (Final research report 1970–1977), 1982. Prepared for the World Zionist Organization – The Settlement Division Jerusalem (in Hebrew, tables with English headings).
2   R. D. Slatyer, Physiological significance of internal water relations in crop

plant, in *Physiological Aspects of Crop Yields*, Symposium published by ASA, CSSA, Madison, Wisconsin, USA, 1969, pp. 53–79.

3   A dunam is equal to 1000 m^2.

4   C. A. Black, Soil Water, in *Soil–Plant Relationships* (New York: John Wiley and Sons, 1957), pp. 39–88; A. Carmi and Z. Plaut, Double Cropping System (Cotton-Wheat) based on frequent drip irrigation and control of the root zone system, in *Optimal Yield Management*, Symposium, Avebury (Aldershot: Gower Publishing Company, 1988), pp. 105–175.

5   U. Or, Drip Fertigation – A case study of the transfer of technology to traditional Arab vegetable growers, in *Optimal Yield Management*, Symposium, Avebury (Aldershot: Gower Publishing Company, 1988), pp. 41–59.

6   F. Bresler, Trickle-drip irrigation principles and application to soil-water management, *Adv. Agron*, 29 (1977): 343.

7   Z. Karchi, Consequences of technological development for traditional family type agriculture in arid regions in Israel. Council of Europe, Parliamentary Conference AS/AGR/conf 15 (1986): 72–78, in European agriculture, 2000, Switzerland.

8   Z. Karchi, Development of melon culture and breeding in Israel (Proc. Cucurbitaceae 2000), *Acta Hort.* 510 (2000): 13–17.

9   Z. Karchi, Consequences of technological development for traditional family type agriculture in arid regions in Israel. Council of Europe, Parliamentary Conference AS/AGR/conf 15 (1986): 72–78, in European agriculture, 2000, Switzerland.

CHAPTER 17

Protected Agriculture: A Regional Solution for Water Scarcity and Production of High-Value Crops in the Jordan Valley

DANIEL J. CANTLIFFE

The Florida/Israeli Protected Agriculture Project (FIPA) was established at the University of Florida <http://www.hos.ufl.edu/protectedag> to foster cooperation between research and extension personnel in Israel and at the University of Florida, as well as commercial companies in both the United States and Israel related to protected agricultural products. Markets around the world for fresh produce have greatly escalated towards consumer demand of high-quality products. Many of these products have been developed in the Jordan Valley region, including Galia-type melons, Beit Alpha cucumbers, and especially technology related to protected agriculture. The FIPA Project has joined researchers in Jordan and Israel to improve technology for sustainability of producing horticultural products in protected structures. One of the most important features related to the FIPA Project is the use of sustainable irrigation water management programs. By controlling the amount of water utilized, both water-use efficiency and circumvention of ground water pollution is a reality. The structures allow for production of fresh vegetables prevalent under the mild winter climates of the Jordan Valley, Israel, and the United States, such as in Florida. Systems are being developed wherein evaporation is minimized and unused water in the irrigation system is recycled for continuous use in the structure. The system relies on the testing of various economical hydroponic or soilless production systems and water recycling systems resulting in the production of high-quality and high-value crops.

There were 150,000 ha of vegetables produced in Florida valued at $1.8

billion during the production season of 1997–98.[1] The major crops of tomato, watermelon, pepper, cucumber, and strawberry accounted for 56% of the total statewide vegetable crop value. Vegetable culture in Florida is a very technological business involving several high-cost inputs including polyethylene mulch, drip irrigation, fertilizer, and pesticides. Currently, almost one-third of Florida vegetables, including all tomatoes, strawberries, peppers, eggplants, and most melons, are produced on polyethylene-mulch. Nearly 50% of the polyethylene-mulched crops are grown with drip irrigation.[2,3]

Although Florida vegetable culture involves intensive production practices, there are major challenges facing the vegetable industry. These challenges are: (1) increased regulation of water, fertilizer, and pesticide inputs; (2) loss of a major soil fumigant, methyl bromide; (3) increased urbanization and loss of some of the more desirable (warmer) production land in southern Florida; (4) continued challenges from weather, including freezes, winds and rain; and (5) competition for water between the agricultural and urban sectors.[4] One may add to these challenges the increasing problems associated with regional and global market competition. The added protection by plasticulture could lead to production of higher quality crops that will make growers more competitive against imports from other vegetable production areas in the world as well as increase water use efficiency. It is evident that for the vegetable industry to prosper and grow, there is a need to develop new cultural technologies.

Plasticulture systems, including greenhouses, could provide a means to deal with the challenges listed above.[5,6,7] Currently, there is a small greenhouse (hydroponic) vegetable industry in Florida, but these special greenhouses represent a substantial investment due to heating and cooling system costs. An alternative might be the use of greenhouse structures with passive ventilation and more effective heating techniques. Greenhouse vegetable culture can provide protection from the weather, a major production challenge faced by vegetable growers. The serious potential loss of crops due to freezes and rain or wind is a major challenge and concern for all vegetable growers in climates such as Florida. These could more easily be controlled in greenhouse culture. Also, greenhouse structures can protect the crop from wind and rain, but also can protect from insects when fitted with insect exclusion screens. Therefore, plasticulture systems could reduce the use of pesticides.

Plasticulture systems could include the use of soilless culture for crop production and thereby increase water use efficiency. One example would be bag or container production using an inert media such as perlite, vermiculite, peat, or coconut fiber. Soilless culture has been used successfully for vegetable production in Florida. Soilless culture would address the current challenges of urbanization because with soilless culture in greenhouses, winter vegetable production would not depend on warm, sandy soils of southern, coastal Florida. In addition, the loss of methyl bromide would be less troublesome if a portion of the vegetables could be grown in soilless culture, either under a protective structure or in open-field soilless culture. Water use efficiency can be improved by recycling "unused" water back to the plants via the fertigation system.

In summary, plasticulture with soilless cultural systems could address several of the serious challenges facing the vegetable industry in Florida and other areas of the world with similar climates. Some of the plasticulture technologies currently exist, but need to be evaluated and refined. Already, this technology is in use in several places in the world, including Israel and other Middle Eastern countries, several Far East countries (China, Korea, Japan), Canada, and Mexico. These countries face some of the same challenges as does the Florida vegetable industry. The Protected Agriculture Project at the Horticultural Sciences Department in Gainesville could provide much needed information for hands-on training and demonstrations so that producers could examine, work, and train in this exciting new agricultural business endeavor.

Protected vegetable production in greenhouses can afford several advantages to producers. They include the ability to moderate temperature during various seasons of the year, wind protection, insect protection, and rain protection. In the past ten-plus years, greenhouse production of vegetables in countries such as Israel has soared. The use of plastic greenhouses, especially for vegetable production, has, simply put, made an oasis out of the desert in many places in Israel. One such example is the proliferation of greenhouse vegetable and flower production in the Arava Desert. In examining production systems in this area, it was conceivable to assume that similar production systems would work in the more humid, semi-tropical areas and countries such as Florida. For that reason, in 1997, the Florida–Israeli Protected Agricultural Project was established in Gainesville, Florida.

Much of the greenhouse production schemes of Israel were brought to Florida to determine if we could economically produce various vegetable commodities in high-roof, passive-ventilated greenhouse structures. In order to do this, we developed a network of Israeli partners, who supplied materials and some resources for us to develop the project as a demonstration research structure in Gainesville. A Top Ltd. greenhouse was constructed in 1999 on approximately 3/10 ha of land. This structure was covered with Ginegar virus-free plastic and the sidewalls were screened against insects with Meteor 50-mesh insect screen. The setup of the greenhouse in this fashion with a vented roof, which was also screened, allowed us to exclude most insects, with the possible exception of thrips, from coming into the greenhouse. It also offered the possibility of reducing potential damage from viruses by restricting reproduction of whiteflies within the greenhouse. By screening the greenhouse, we were also given the opportunity to use bumble bees as pollinators for melons, peppers, and tomatoes.

During our first year of production, we conducted variety trials on tomato, pepper, cucumber, and melon. We also investigated the effects of plant density and shoot pruning on yield and quality of sweet peppers produced in the summer.[8,9] A disorder on sweet pepper known as 'Elephant's Foot', which is seen when peppers are produced in hydroponic culture using media such as perlite, was observed and circumvented by technology developed via the Project.[10] Our variety trials consisted of germplasm from both Hazera and Zeraim Seed Companies of Israel, and several cultivars of the various commodities, which were provided as checks from Dutch seed companies. Tomatoes,[11] peppers, melons,[12] and cucumbers[13] were planted during 1999–2001, and the production schemes for the cultivar trials were the use of bag or plastic pot culture and perlite or pine bark as a media. Perlite bags were 1 meter in length. Plants were planted at 0.4 meters apart and in single rows for all crops except tomatoes, which were planted in double rows. All plants were fertigated at each irrigation on a timed basis as related to sunlight and temperature within the house. Irrigation frequency was regulated by drainage frequency wherein drainage was generally maintained at less than 35% of water applied. All plants were permitted to grow in a vertical fashion to guide wires across the center of the greenhouse, approximately 4 meters high. Harvesting of all commodities was done either at full slip for melons, full color for tomato and peppers (red/yellow), or economic maturity for cucumbers. For all but the Beit Alpha cucumbers, bumble bees were used for pollination. Insect pests were monitored daily and controlled by beneficial insects and when absolutely necessary by approved biological and/or chemical pesticides. Sticky traps were used to collect and identify potential insect pests.

In other experiments that tested the plant density and pruning methods for peppers, plant populations of 2, 3, and 4 plants per sq. m. and 66.5, 43.3, and 33.3 cm (in-row spacing) and shoot pruning methods of 1-, 2-, and 4-stems were examined using the pepper cultivar Robusta.[8,9] Marketable yield, both for number and for weight per meter squared, increased linearly with plant density and were greater on plants with 4 stems than those with 2 or 1 stem. Density had no effect on production of extra large fruit. Total marketable yield and extra large fruit yield per plant were greatest in the 4-stem plants at 2-plants/sq. m. The results of these studies indicated that 12-plants/sq. m. pruned to 4 stems led to increased marketable and extra large fruit yield in a short harvest period of the summer greenhouse pepper crop grown under mild winter climate conditions. Subsequent trials to this have improved yields and the quality of the crop and reduced labor by changing the pruning system from the Dutch-type to the Spanish/Israeli-type.

The development of irrigation management strategies that lead to high marketable fruit yields while using small amounts of water and fertilizer and that lead to reduced incidences of irrigation-related fruit disorders is much needed in order to make recommendations on irrigation practices for crops grown in soilless media. Recently we determined the effects of three types of media under five irrigation schedules and two volumes of nutrient solutions per irrigation event on pepper plant growth, fruit yield and quality, and the efficiency of water and fertilizer usage for marketable fruit production of plants grown in a mild winter climate. Five irrigation schedules were created by starting irrigation events at different levels of solar radiation integrals. The number of irrigation events per day varied with daily and seasonal climate changes. In addition, either one volume or a double volume of nutrient solution was delivered to the plants at each irrigation event. Both of these volumes of nutrient solutions delivered the same amount of nutrients per irrigation event. A second and simultaneous experiment evaluated fruit yield and quality in plants under the same five irrigation schedules and three types of media but with double amounts of water and fertilizer at each irrigation event.

Similar marketable fruit yields were obtained from plants irrigated with one volume per event at high irrigation frequencies and with double the volume per event at high and low frequencies. Although good fruit yields were obtained with the high number of irrigation events per day, such frequent irrigation led to high amounts of water and fertilizers used and to a high percentage of fruit with cracking. However, it was possible to identify irrigation treatments that led to high fruit yields with a low incidence of cracking and with low use of water and fertilizer. Plants grown in any of the three media, irrigated with 12 or 16 events per day and double the volumes per event, yielded 9.0 and 9.6 $kg \cdot m^{-2}$, respectively, with 58% and 44% less water, and with 80% and 73% less fertilizer used, respectively, compared to irrigating with 62 events per day at the lower volume per event. These results indicated that it was possible to produce about 9 $kg \cdot m^{-2}$ of fruit under low cost greenhouse structures and with low heating during winter, and that irrigation could be managed to minimize water and fertilizer use, as well as fruit disorders, without decreasing fruit yield.

The twenty-first century brings more people, less water, more demand for world food production, and a sign of hope for the future. Vegetable agriculture with its importance for human nutrition has gone through many production changes in the past 100 years. Science has taught farmers how to intensify their efforts many fold, giving them at present the luxury, and curse, to over-produce. Unfortunately, as world economies dramatically improve, demands for land for nonagricultural use has likewise dramatically increased. Many science-based alternatives to insure high productivity have been diminished including the dependency on methyl bromide as a plant bed sterilant. The result is a scramble for export economies to drastically change vegetable production schemes. Protected agricultural systems in warm winter climates will surpass much of the open field production of today. Alternatives for soil-based systems as well as improved pest management are current problems facing such protected agricultural production schemes. Breeding programs to maximize efficiency of such protected agricultural systems are likewise essential. Plant growing structures must conform to the needs of plant productivity, as well as production and water economics. Most importantly, field production-based agricultural systems of such places in North America, Mexico, the Middle East, etc., must be prepared to change to more intense protected agricultural systems in as little as the next 5 years. The future for efficient economic vegetable production on a year-round basis will be dependent on these science-based changes.

Summary

Florida produces $1.8 billion of vegetables on 160,000 ha of land. All of this production is destined for the fresh market and most of the produce is shipped to Northern United States markets. Most of this vegetable production is grown in the field out of season in the winter months, thus requiring land not prone to freezes. Unfortunately, Florida is becoming highly urbanized with the population exceeding 15.3 million in 2000. The major impact of urbanization has been a loss of Florida's warmest and most productive lands for winter vegetable production. Competition between the agricultural and urban sectors for water has created a need to improve water use efficiency for crop production. The use of protected structures can reduce water requirements for freeze protection in winter and greatly increase water use efficiency during hot months by recycling unused irrigation water within the structure. In 1997, a Florida and Israeli Protected Agriculture Project was initiated in order to take better advantage of land distal from the urbanized coastline. An 8-m high passive-ventilated Israeli-style greenhouse was constructed in north Florida, a minimum of 85 km from either coast. Successful pepper, tomato, cucumber and muskmelon crops were grown as fall-winter and spring-summer crops. With proper shading, heat-sensitive crops could be produced throughout the summer. Moreover, a class of high quality vegetable crops which could not be produced under typical field conditions in Florida's climate were produced. These included Galia-type muskmelon, Beit alpha cucumber, cluster tomato, and high-quality colored peppers. Yields from greenhouse crops are generally 10 times more than comparable field-produced crops. For further information regarding the Florida/Israeli Protected Agriculture Project, please visit the website http://www.hos.ufl.edu/protectedag.

Notes

1 J. D. Witzig, *Florida Agricultural Statistics – Vegetable Summary 1997–98* (Florida Agric. Statistics Serv., 1999).

2 G. J. Hochmuth, D. J. Cantliffe, Z. Karchi, and I. Secker, The Florida Center for Plasticulture, *Proceedings 27th National Agricicultural Plastics Congress*, American Society of Plasticulture (1998), pp. 231–36.

3 G. J. Hochmuth, D. J. Cantliffe, Z. Karchi, and I. Secker, The Florida Center for Plasticulture, *Plasticulture* 118 (1999): 58–66.

4 D. J. Cantliffe, E. Jovicich, and G. J. Hochmuth, Where has all the good land gone? Protected vegetable culture – our future, *Greenhouse Techniques Toward the Third Millennium*, Haifa, Israel, 1999.

5 E. A. Waldo, G. J. Hochmuth, D. J. Cantliffe, and S. A. Sargent, Protected Winter Production of 'Galia' Muskmelons, *Proceedings of the Florida State Horticultural Society* 110 (1997): 303–5.

6 E. A. Waldo, G. J. Hochmuth, D. J. Cantliffe, and S. A. Sargent, Growing 'Galia' Muskmelons using Walk-in Tunnels and a Soilless Culture System in Florida and the Economic Feasibility of using the Systems, *Proceedings of the Florida State Horticultural Society* 111 (1998): 62–69.

7 E. A. Waldo, G. J. Hochmuth, D. J. Cantliffe, and S. A. Sargent, Technical and Economic Feasibility of Growing 'Galia' Muskmelons in the Winter in Northern Florida Using Protective Structures and Soilless Culture, Abstract. *Proceedings 28th National Agricultural Plastics Congress* (1999), pp. 115.

8 E. Jovicich, D. J. Cantliffe, and G. J. Hochmuth, Plant Density and Shoot Pruning on Yield and Quality of a Summer Greenhouse Sweet Pepper Crop, *HortScience, Abstract* 34 (1999): 532.

9 E. Jovicich, D. J. Cantliffe, and G. J. Hochmuth, Plant Density and Shoot Pruning on Yield and Quality of a Summer Greenhouse Sweet Pepper Crop in North Central Florida, *Proceedings 28th National Agricultural Plastics Congress* (1999), pp. 184–90.

10 E. Jovicich, D. J. Cantliffe, and G. J. Hochmuth, 'Elephant's Foot': a Plant Disorder in Hydroponic Greenhouse Sweet Pepper, *Proceedings of the Florida State Horticultural Society* 112 (1999): 310.

11 J. C. Rodriguez, D. J. Cantliffe, and N. Shaw, Performance of greenhouse tomato varieties grown in soilless culture in north central Florida, *Proceedings of the Florida State Horticultural Society* 114 (2001): 303–6.

12 N. L. Shaw, D. J. Cantliffe, and S. Taylor, Hydroponically produced 'Galia' muskmelon – What's the secret? *Proceedings of the Florida State Horticultural Society* 114 (2001): 288–93.

13 N. L. Shaw, D. J. Cantliffe, J. C. Rodriguez, S. Taylor, and D. Spencer, Beit Alpha Cucumber – An Exciting New Greenhouse Crop, *Proceedings of the Florida State Horticultural Society* 113 (2000): 247–53.

CHAPTER 18

Postscript: Focusing on Peace – Building Trust and Understanding

JOSEPH GINAT AND MATTHEW CHUMCHAL

The Center for Peace Studies (CPS) seeks to foster peace between groups in conflict. Although governments lay the cornerstone by signing treaties, it is the citizens of a nation that build peace. To reach these citizens, CPS organizes conferences that focus on issues important to groups on both sides of a conflict. Although the proximate goal of these conferences is to generate solutions to a specific problem, the ultimate goal is to build trust and understanding among participants. The friendships formed at these conferences help build networks of trust between the groups in conflict, allowing the enemies of today to become the friends of tomorrow. Previous issues addressed by CPS conferences include the Middle East peace process and Palestinian refugees. Although it is often overshadowed by high profile issues, like the Palestinian refugees, the issue of water is paramount and was the focus of a 2001 CPS conference and this book.

Water is the source of life. The availability of water has been central to the development of society. Due to the presence of large rivers, some regions of the Middle East are relatively water rich. It was in these fertile regions that the earliest civilizations developed. The ancient Greeks called the valley between the Tigris and the Euphrates Mesopotamia, the land "between the rivers." Years of enriching silt deposits from the rivers and irrigation canals made this land highly productive. Similarly, the area around the Nile River is lush and green, but the land becomes increasingly arid the farther one travels from the life-giving river. As with Mesopotamia, floods deposited enriched soil each year, making the Nile delta the "breadbasket" of the Mediterranean basin in ancient times.

In the Nile delta, agricultural production results from the farmers' diligence. By digging canals from the Nile to areas adjacent to the river, farmers increase the amount of arable land. However, in many areas of the

Middle East there are no large rivers that can be utilized during times of drought. The Bible tells us that, unlike Egypt, the region occupied by present-day Israel, Palestine, and Jordan was dependent on rain.

> For the land, whither thou goest in to possess it, is not as the land of Egypt, from whence ye came out, where thou sowedst thy seed, and wateredst it with thy foot, as a garden of herbs: But the land, whither ye go to possess it, is a land of hills and valleys, and drinketh water of the rain of heaven."[1]

As in ancient times, drought is common in this arid region. Even in years when rain is plentiful there is a drawdown in the water-level of Lake Kinneret (Lake Tiberius, Sea of Galilee) as well as the regional aquifers, the sources of water upon which Israel, Jordan, and Palestine depend.

The reliance on shared water resources, and the shortage of water in the Israel–Jordan–Palestine region, complicates the tenuous political situation. After the final Israeli–Palestinian peace agreement, refugees will return to Palestine and water shortages will be exacerbated. Although water issues are not politicized and rarely find their way into the headlines, they are just as important to the peace process as the fate of the Temple Mount and the location of Israel's borders. Thus, water is a principal issue in both the Israeli–Jordanian peace treaty and the Oslo Accords signed by Israel and the Palestinian Authority. In fact, water is one of the issues yet to be resolved in the permanent status negotiations between Israel and the Palestinians.[2]

Israel, Jordan, and Palestine must cooperate to ensure that there will be adequate supplies of clean, fresh water. The task will not be easy. Although there are precedents for the joint management of international rivers (e.g. the Danube River in Europe), there are no cases of international management of aquifers. Israel, Jordan, and Palestine must work together to regulate the drilling of new wells and to prevent over-pumping. In addition, the treatment and discharge of sewage that can contaminate the regional aquifers must be properly managed. Successful joint management will also require each country to participate in the collection and sharing of data.

As the population of the Middle East and the demand for water grows, cooperation between Israel, Jordan, and the Palestinian Authority must be expanded to include other countries in the region including Iraq, Lebanon, Syria, and Turkey. Not only will additional sources of water need to be developed but there will be a greater need for joint management of shared water resources. For example, Lake Kinneret supplies half of Israel's drinking water and smaller amounts to Jordan and Palestine via bi-national agreements. The source of Lake Kinneret, the Jordan River, is fed by three tributaries, only one of which, the Dan River, is within Israel's borders. Headwaters of the two other tributaries, the Banias and the Hazbani, are located in Syria and Lebanon, respectively, two countries that are presently not at peace with Israel. Changes in flow in the Banias or

Hazbani could affect the Jordan River, which in turn could affect the water budgets of Israel, Jordan, and Palestine. Similarly, alteration of the Tigris and Euphrates systems could affect the water budgets of Syria, Turkey, and Iraq.

Water shortages are not confined to the Middle East; the global water situation is becoming increasingly more alarming.[3] Worldwide, growing population centers are exerting a greater demand on fresh water supplies, while at the same time the quality of the remaining fresh water supply is threatened by pollution. The water crisis in the Middle East and the rest of the world can be alleviated. However, ending the water crisis will require a greater effort to collect and share data, the development and implementation of technological solutions, and international cooperation. The 2001 conference on water in the Middle East was one step toward seeking solutions to the water crisis.

The Center for Peace Studies, with the financial assistance of the Citizens Exchange program of the US Department of State, has been able to continue the momentum that began at the 2001 conference. In the summer of 2003, water working groups consisting of water experts, community and political leaders, and students (both from the region and the US) were established. Because of the political realities, two sets, a Northern and Southern tier, worked in parallel. The Northern Tier consisted of Iraqis, Jordanians, Lebanese, Syrians, Turks, and Americans, while the Southern Tier included Israelis, Jordanians, Palestinians, and Americans. The Jordanians acted as a "bridge" between the groups, while the Americans acted as facilitators. Like the conference participants, the participants in the water working groups represented a variety of disciplines. They focused on the technical details of alleviating the water crisis and used the meetings as opportunities for trust-building and for establishing networks of communication.

The Southern Tier water working group met four times in Cyprus and Jordan. During these meetings, participants agreed that water shortages are a regional problem and that all states in the region, not just those in the Southern Tier, must cooperate to develop solutions. Technological solutions, public education programs, and water conservation at the level of the individual and community were identified as essential components of the solution to the water crisis. The participants proposed the implementation of several projects through partnerships with government agencies, academic institutions, businesses, and non-governmental organizations (NGOs). The creation of water efficient model villages was proposed as a method of developing, implementing, and demonstrating the technological, educational, and conservation ideas discussed at the meetings. These model villages would be located in each of the Southern Tier countries and set the standard for access to water and conservation.

Participants agreed that joint management of water resources will only be possible in an atmosphere of transparency. The participants proposed a regional water database as a means to enhance cooperation and facilitate trust on water issues. The database would include consumption rates, data on source water quantity and quality, information on regional hydrology and meteorology, information on treatment facilities, and a directory of regional water experts.

Finally, participants proposed a workshop to develop a public education campaign focusing on water conservation. The core aspects of the campaign would be developed during the workshop and representatives from each country would then refine the campaign to meet the unique needs of their country. The proposed campaign would include school curricula, public lectures by water experts, and media advertisements. The campaigns would target primary and secondary schools, universities, and communities.

To "jump start" these and other projects, as well as to facilitate cooperation between Southern Tier countries, participants proposed the development of collaborative research projects involving doctoral students from each of the Southern Tier countries. The Center for Peace Studies has taken steps to secure funding and resources for these students through the University of Oklahoma.

The Northern Tier water working group met six times in Lebanon, Syria, and Turkey. As in the Southern Tier meetings, Northern Tier participants agreed that regional cooperation is essential in order to alleviate water shortages. Indeed, after the first meeting, they made "Regional Cooperation for Water Management" the theme of subsequent meetings. Northern Tier participants emphasized the need for compiling a regional water database and developing the region's data collection infrastructure. Not only will compiling all available data help to facilitate trust and cooperation, but it will help regional experts identify data collection priorities.

The Northern Tier meetings led to a development initiative between Iraqis, Syrians, Turks, and Americans known as the Euphrates-Tigris Initiative for Cooperation (ETIC). The goals of ETIC are to improve the quality of life and promote harmony between people living in the Euphrates-Tigris region. ETIC will achieve these goals by promoting dialogue about development issues and initiating projects that benefit all groups in the region. ETICs founders hope that this initiative will serve as a model for development projects throughout the Middle East.

Graduate student participants from the Northern and Southern Tier meetings expressed hope that the friendships formed at the water working group meetings will lead to continued dialogue and international collaborations. Student participants have kept in contact with one another via email "newsletters" and a website. The newsletters and website feature the professional accomplishments of water working group students and news about water working group projects and proposals. In addition, the website provides links to pictures and reports from water working group meetings.[4]

Participants from both the Northern and Southern Tier meetings agreed that the water working groups are an important part of the solution to the Middle East water crisis. Ultimately, alleviating water shortages in the Middle East will require a commitment to pursue cooperation from all parties. The water working groups helped to build trust and facilitate cooperation between water experts. It is our hope that the collaborative relationships formed between water working group participants will lead to improvements in the regional water situation, which, in turn, will make the larger goal of peace a reality.

Notes

1   The Holy Bible, Deuteronomy 11: 10–15.
2   E. Feitelson and M. Haddad, Introduction, in: E. Feitelson and M. Haddad (eds.), *Management of Shared Ground Water Resources. The Israeli–Palestinian Case with an International Perspective* (Boston: Kluwer Academic Publishers, 2000), p. xiii.
3   P. H. Gleick. The world's water: The biennial report on freshwater resources, 2000/2001 (Washington, D.C.: Island Press, 2001).
4   A link to the student website is located on CPS's web site: <http://www.ou.edu/ipc/cps/>.

Contributors

Alfred Abed Rabbo is an associate professor at Bethlehem University. He is a specialist in environmental chemistry with a particular interest in water science. He founded the Water & Soil Environmental Research Unit (WSERU) at Bethlehem University in 1989 and is its current director. He has co-authored three books: *Water Soil and Plant Quality-Testing Procedures*; *Springs in the West Bank: Water Quality and Chemistry*; and *Wells in the West Bank: Water Quality and Chemistry*; and over twenty scientific publications, mostly dealing with water issues in the Palestinian West Bank. His research interest is in environmental issues, mainly water issues, particularly water quality, and water resource studies including hydrology and hydrochemistry. He is also involved in water awareness programs concerned with capacity building in water resource management, and conducts research on experimental waste water treatment technologies, and aquifer modeling.

Mohammed Issa Taha Ali is an assistant professor at Princess Sumaya University for Technology in Amman, Jordan, where he has taught since 1992. He previously was the head of the Economics Department at the Royal Scientific Society. He also served as the head of Industrial Economics Division, and was an economic researcher at the Royal Scientific Society. He has his PhD in econometrics from the University of Leeds in the United Kingdom and is the author of books, working papers and articles in professional journals. Dr. Ali's research interests include regional economic cooperation and econometric modeling.

Rateb Mohammad Amro is director general and founder of Horizon Center for Studies and Research in Amman, Jordan. He was director of the Occupied Territories Office of the Ministry of Foreign Affairs and is a retired colonel of the Jordan Army Forces. He is the author of articles and chapters of books, including "A Jordan Perspective" in Joseph Ginat and Edward Perkins (eds.) *The Palestinian Refugee Problem: Old Problems – New Solutions* (Norman, OK: University of Oklahoma Press, 2001; Brighton: Sussex Academic Press, 2001) and "The Peace Process: A

Jordanian Perspective" in Joseph Ginat, Edward J. Perkins, and Edwin G. Corr, *The Middle East Process: Vision Versus Reality* (Norman, OK: The University of Oklahoma Press, 2002; Brighton: Sussex Academic Press, 2002).

Yaakov Anker is a PhD student in the Department of Geophysics and Planetary Sciences at Tel Aviv University, where he participates in the multi-lateral working group of German, Israeli, Jordanian, and Palestinian scientists studying the hydrogeology of the Jordan Valley.

David L. Boren became the thirteenth president of the University of Oklahoma in 1994 and is also a professor of political science there. Prior to coming to the University, he served in the US Congress as a senator from Oklahoma and earlier as governor of Oklahoma, the first person in history to hold all three positions. President Boren is widely respected for his long-time support of education, his distinguished career as a reformer of the American political system, and his innovations as a university president. A graduate of Yale University in 1963, he majored in American history, graduated in the top one percent of his class, and was elected to Phi Beta Kappa. He was selected as a Rhodes Scholar and earned a master's degree in politics, philosophy, and economics from Oxford University in 1965. In 1968, he received a law degree from the University of Oklahoma College of Law, where he was on the *Law Review*, elected to the Order of the Coif, and won the Bledsoe Prize as the outstanding graduate by a vote of the faculty. As a United States senator, Boren was the longest-serving chairman of the Senate Select Committee on Intelligence. Boren has served as a member of the Yale University board of trustees and as chairman of the Department of Political Science and chairman of the Division of Social Sciences at Oklahoma Baptist University. He is the co-editor of two books and the author of published articles and book chapters.

Daniel J. Cantliffe is a professor and chair of the Horticultural Sciences Department at the University of Florida. He received his PhD from Purdue University where he specialized in plant physiology and plant nutrition. Dr. Cantliffe's research is currently focusing on greenhouse crop production. He has published over 700 papers, 175 of which are peer-reviewed.

Matthew Chumchal is a graduate student in the Department of Zoology at the University of Oklahoma. Mr. Chumchal is studying water pollution and its relationship to fisheries management and attended two of the Middle East Water Working Group Meetings as the American student observer.

Lea Davidson is a researcher in the Department of Geophysics and

Planetary Sciences at Tel Aviv University since 1997. She received her PhD in Hydrochemistry from Moscow Institute of Global Climate and Ecology. Dr. Davidson's studies are on groundwater salination and sources of contamination in multi-aquifer systems.

Eran Feitelson is an associate professor in the Department of Geography and the head of the School of Public Policy of the Hebrew University of Jerusalem. He holds a PhD from Johns Hopkins University. Dr. Feitelson is also the chair of the Israeli National Parks and Nature Reserves Board. His academic work focuses on environmental policy issues, especially management of trans-boundary water resources.

Franklin M. Fisher is the Jane Berkowitz Carlton and Dennis William Carlton Professor of Economics, Emeritus at the Massachusetts Institute of Technology, where he taught for 44 years. He serves as the chair of the Water Economics Project, a cooperative endeavor of American, Dutch, Israeli, Jordanian, and Palestinian experts that is facilitated by the government of The Netherlands with the knowledge of the regional governments. Professor Fisher is the senior author of *Liquid Assets: An Economic Approach for Water Management and Conflict Resolution in the Middle East and Beyond* and the author or co-author of over 150 articles and 16 other books including *The Identification Problem in Econometrics, Folded, Spindled, and Mutilated: Economic Analysis and U.S. v. IBM,* and *Disequilibrium Foundations of Equilibrium Economics.*

Akiva Flexer is a professor in the Department of Geophysics and Planetary Sciences at Tel Aviv University where he served as chair during 1997–2001. Dr. Flexer currently heads a Tel Aviv multi-lateral working group of German, Israeli, Jordanian and Palestinian scientists studying the hydrogeology of the Jordan Valley.

Joseph Ginat is a professor emeritus of social and cultural anthropology at Haifa University, a visiting professor of anthropology and the deputy director of the Center for Peace Studies at the University of Oklahoma, and the vice president for international relations and research and director of the Strategic Dialogue Center of Netanya University. He is former chairman of the Israeli Academic Center in Cairo and a former chairman of the Jewish–Arab Center at Haifa University. He is author of many books and articles on the Middle East, including *Blood Revenge: Family Honor, Mediation and Outcasting* (2nd edition) (Brighton & Portland: Sussex Academic Press, 1997); with I. Altman, *Polygamous Families in Contemporary Society: Coping with Challenging Life Style* (Cambridge: Cambridge University Press, 1996); with Edward J. Perkins and Edwin G. Corr, *The Middle East Peace Process: Vision Versus Reality* (Norman, OK:

The University of Oklahoma Press; and Brighton & Portland: Sussex Academic Press, 2002).

Joseph Guttman is chief hydrologist for Mekorot, the Israeli national water company. He is responsible for all geology and hydrology issues in Mekorot and hydrology advisor to the Israeli Government on groundwater issues with the neighboring countries (especially with the Palestinian Authority). He received his PhD and post doctorate in hydrology from Tel Aviv University. He is a former chairman of the Israel Association for Water Resources (IAWR).

Munther J. Haddadin is a consultant based in Amman, Jordan, a courtesy professor at Oregon State University, and an affiliate professor at the University of Oklahoma. Formerly Dr. Haddadin served as minister of water and irrigation of Jordan (1997–8); as chief Jordanian negotiator over water, energy and the environment in the Middle East Peace Process (1991–5); consultant (1987–97); president and chairman of the board of the Jordan Valley Authority (1982–7), and its senior vice president (1973–82). Dr. Haddadin is the author o f *Diplomacy on the Jordan* (Kluwer Academic Publishers, 2001), co-author of *The Inside Story of Peace Making* (UK: Garnet Press, 2005), and several articles on water issues published in professional journals. He was selected as the lecturer for the "Howell Second Century Lectureship" by the International Programs Center of the University of Oklahoma, 2003; and chosen as author of the Best Research Paper Award by the International Water Resources Association, 2001, and by the International Water Resources Association of Oklahoma for the year 2003.

K. David Hambright is an assistant professor in the University of Oklahoma Biological Station and Department of Zoology. He received his PhD in ecology and evolutionary biology from Cornell University. Dr. Hambright is also an adjunct research scientist at the Yigal Allon Kinneret Limnological Laboratory in Israel, where for 10 years he conducted research on various aspects of limnology (the study of inland waters) relating to water quality in the Sea of Galilee, such as plankton and fish ecology, eutrophication abatement and lake rehabilitation. He has trained numerous undergraduate, graduate and post-doctoral students and published more than 50 research articles on various aspects of limnology. Through the University of Oklahoma's Center for Peace Studies, he has been active in leading regional discussions and bridging partnerships among water experts in the Middle East.

His Royal Highness Prince El Hassan bin Talal was born the son of King Talal bin Abdullah and Queen Zien El Sharaf, and is the younger brother of his late majesty King Hussein. Prince El Hassan earned from Christ Church, Oxford University in England a BA with honors in oriental studies, followed by an MA. He has been awarded many honorary degrees and prizes from around the world. His Royal Highness was officially invested as Crown Prince to the Hashemite Throne of Jordan in 1965 and served as King Hussein's closest political adviser, confidant and deputy, as well as acting as Regent when the King was absent from the country, until King Hussein's death. The Prince has initiated, founded and is actively involved in numerous Jordanian and international institutes and committees. Among his multiple prestigious international activities, he is the co-chair of the UN Independent Commission on International Humanitarian issues. He serves on the boards of many other UN organizations. His Royal Highness has honored the Center for Peace Studies by serving as chairman of the board of advisors. The Center's institutional members are the Jordanian Horizon Center for Studies and Research, Haifa University, Bethlehem University, and the University of Oklahoma. He serves as one of three presidents of the Strategic Dialogue Center of Netanya Academic College. In the field of religion his contacts and meetings have evolved into a systematic interfaith dialogue among Muslims, Christians, and Jews; and in Jordan, he set up the Royal Institute for Inter-Faith Studies in 1994. He has promoted sports within the Kingdom. Prince El Hassan is the author of five books and has contributed to newspapers, and magazines. He is fluent in Arabic, English, and French, has studied Biblical Hebrew, and has a working knowledge of German, Spanish, and Turkish.

Zvi Karchi is a former deputy director for research and development of the Israeli Agricultural Research Organization (Volcani), founder and director of the Jordan Valley Research and Development Authority, and a senior scientist in charge of the National Cucurbits Genetics and Breeding Section. He holds an MSc from the University of California and a PhD in plant genetics from the University of Minnesota. Dr. Karchi's research has been largely devoted to the genetics and breeding of muskmelons and watermelons. He has breeder's rights of about 27 F_1 hybrids and cultivars, including the internationally marketed 'Galia' F_1 hybrid. He has published about 200 papers on both field crops and cucurbits. He developed advanced technologies to optimize yield, produce quality, and under desert conditions maximize income per unit area for farmers.

Yoav Kislev is a professor (emeritus) in the Department of Agricultural Economics and Management at the Hebrew University of Jerusalem. He received his PhD in economics from the University of Chicago. Dr. Kislev's research focuses on agricultural economics, the economics of agricultural research, agricultural cooperatives and water. He has served as consultant to the World Bank and is former director of research for Israel's Center for Agricultural Economic Research (Rehovot).

Moshe Ma'oz is professor of Islamic and Middle Eastern studies and former director of the Truman Peace Research Institute at the Hebrew University of Jerusalem. He is the senior deputy director of the Strategic Dialogue Center of Netanya Academic College. He was a Senior Fellow of the Jennings Randolph Fellowship Program at the US Institute of Peace. Dr. Ma'oz is a leading expert on Syria and was assistant advisor on Arab issues to Prime Minister Ben-Gurion, advisor on Arab affairs to defense minister Ezer Weizman, and a member of the Advisory Committees on Arab–Israeli Relations to prime ministers Shimon Perez and Yitzhak Rabin. He is author and editor of 15 books and many articles on Arab–Israeli relations, political and social history of Syria and Palestine, and religious and ethnic communities, including *Assad, the Sphinx of Damascus: A Political Biography* (London and New York: Weidenfeld and Nicolson, 1998).

Mohammad Abudayyeh Matouq is president of Jordan Japan Academic Society (JJAS). He has served as dean of Maan College, head of research and development at the National Training of Trainers Center, and he is currently on the academic staff at the Faculty of Engineering Technology, at Al-Balqa Applied University (BAU) in Jordan. He received his PhD in chemical engineering from Nagoya University, Japan. Previously, Dr. Matouq served as a senior engineer at the Jordanian Ministry of Energy and Mineral Resources and worked at the United Nations Centre for Regional Development – Environment and Human Resources Development.

F. Jamil Ragep is a professor of the history of science at the University of Oklahoma and was until recently Co-Director of the Center for Peace Studies and Coordinator of Middle Eastern Studies. He received his undergraduate and master's degrees from the University of Michigan and his PhD from Harvard University. In addition to his research on science in Islamic civilization, for which he has been awarded numerous fellowships and designated the winner of the prestigious Kuwait Prize, he coordinated in 2003 and 2004 multinational working groups dealing with water in the Middle East, a project funded by the US State Department. Among his publications is a major study dealing with Islamic astronomy *Nasir al-Din al-Tusi's Memoir on Astronomy* (New York: Springer-Verlag, 1993) and a collection of essays (co-edited with S. P. Ragep) on the transmission of science between cultures *Tradition, Transmission, Transformation* (Leiden: E. J. Brill, 1996).

Eliyahu Rosenthal is a professor of hydrogeology in the Department of Geophysics and Planetary Sciences at Tel Aviv University. Formerly, Dr. Rosenthal was senior scientist at the Israel Hydrological Service, acted as coordinator with the Jordanian authorities for the joint exploitation of water resources in the Jordan Valley, and negotiated water issues within the framework of the Israeli–Jordanian Peace Treaty. Most of his scientific work deals with multiple aquifer systems and with groundwater salinization processes, particularly hydrochemical interaction between brines and fresh groundwater.

David J. Scarpa has been teaching earth and environmental sciences at Bethlehem University since 1990. He is chief scientist to the Water and Soil Environmental Research Unit, part of the University's Outreach Program to the local community. He was appointed dean of science in 2004. He has a BSc Honours Degree in earth sciences from the Open University of the United Kingdom, an MSc in geology from Liverpool University, and a PhD in hydrology from London University. He has authored or co-authored three books and more than 30 internationally refereed scientific papers.

Aondover Tarhule is an associate professor of geography at the University of Oklahoma. He holds a PhD from McMaster University, Canada. Prior to joining the University of Oklahoma, Dr. Tarhule taught at the University of Jos in his native Nigeria, and did post-doctoral research at Queens University, Canada. His areas of research focus include physical hydrology as well as hydroclimatic variability and its impacts on society. A significant proportion of his work is based on West Africa where he has traveled extensively and has major on-going research exploring the use of tree rings as climate proxies.

Onn Winckler is a senior lecturer in the Department of Middle Eastern History and the head of the Department of Multidisciplinary Studies at the University of Haifa. His research specializes in the demographic and economic history of the modern Middle East.

Annat Yellin-Dror is the senior researcher of geology and geophysics and general director of the Hydrogeological Information Center (GIC) at The Department of Geophysics and Planetary Sciences, Raymond and Beverly Sackler Faculty of Exact Sciences, Tel-Aviv University. He earned a BSc in geology at Ben-Gurion University (1979) MSc and PhD in geophysics at the Department of Geophysics and Planetary Sciences, Tel Aviv University. His research fields of interest are geophysical log interpretation, basic analysis and data systems. He has been a senior investigator in international research projects conducted in the United States, Italy,

Palestine and Jordan. Dr. Yellin-Dror served on the Board of Directors of the Israel National Oil Company and The Environmental Company of Israel.

Tamar Zohary is a senior research scientist at the Yigal Allon Kinneret Limnological Laboratory in Israel and a visiting senior research scientist at the University of Oklahoma Biological Station. She received her PhD from the University of Natal in South Africa. Her research centers on the ecology of phytoplankton (microscopic algae) and other microorganisms in freshwater and marine ecosystems, and on ecological modeling of lake ecosystem functioning.

Index